Lecture Notes in Earth Sciences

47

Editors:
S. Bhattacharji, Brooklyn
G. M. Friedman, Brooklyn and Troy
H. J. Neugebauer, Bonn
A. Seilacher, Tuebingen

Ralf Littke

Deposition, Diagenesis and Weathering of Organic Matter-Rich Sediments

Springer-Verlag
Berlin Heidelberg GmbH

Author

Dr. Ralf Littke
Forschungszentrum Jülich GmbH
Postfach 19 13
W-5170 Jülich, FRG

"For all Lecture Notes in Earth Sciences published till now please see final pages of the book"

ISBN 978-3-540-56661-8

Library of Congress Cataloging-in-Publication Data

Littke, Ralf, 1957
Deposition, diagenesis, and weathering of organic matter-rich sediments/Ralf Littke. p. cm. - (Lecture notes in earth sciences; 47)
Includes bibliographical references.
ISBN 978-3-540-56661-8   ISBN 978-3-540-47622-1 (eBook)
DOI 10.1007/978-3-540-47622-1

1. Sediments (Geology) 2. Organic geochemistry. 3. Petroleum-Geology. I. Title. II. Series.
QE471.2.L55 1994 551.3-dc20 93-17490 CIP

© Springer-Verlag Berlin Heidelberg 1993

Originally published by Springer-Verlag Berlin Heidelberg New York in 1993

Typesetting: Camera ready by author
32/3140-543210 - Printed on acid-free paper

PREFACE

The present interest in sediments which are rich in organic matter results not only from their economic significance as potential oil and gas source rocks, but also from the fact that their deposition is the result of special environments. Subtle changes in the environmental conditions may result in great variations in the geochemical and petrographical characteristics of the organic matter. Therefore, the study of organic matter-rich sediments can provide a key to past sedimentary conditions. In addition, the elucidation of the depositional controls is of importance for oil and gas exploration strategies, for which the knowledge of source rock distribution and quality is critical. Furthermore, organic matter reacts extremely sensitive to changes in temperature during burial. The result of this sensitivity is the generation of volatile products such as carbon dioxide, water, nitrogen, oil and gas and a reorganization of the solid organic residue. Some of these changes are quantified as maturity parameters which can be used as calibration tools in basin modelling, i.e., in the modelling of temperature histories of sedimentary basins. The use of maturity parameters and other organic matter characteristics as indicators for diagenetic conditions and depositional processes is, however, restricted, if analyses are performed on outcrop samples, because weathering also affects organic matter.

The aim of this book is to provide information on deposition, diagenesis and weathering of organic matter-rich sediments by assembling general information and by presenting case studies. The text was originally written as "Habilitationsschrift" and is a summary of seven years of research work performed at the Institute of Petroleum and Organic Geochemistry (ICG-4)" at the Federal Research Centre Jülich (KFA).

Although this summary is a monography, it greatly benefitted from the cooperation with numerous scientists inside and outside this institute and from the help of the technical staff of ICG-4. Especially, I want to thank D.H. WELTE, director of ICG-4, for permanent support, continuous interest and many important stimulations. Also, I am very grateful to D.R. BAKER, Houston, J. RULLKÖTTER, Oldenburg, and R.F. SACHSENHOFER, Leoben, who read an earlier version of this text and gave valuable advice, to D.K. RICHTER, Bochum and B. STÖCKKERT, Bochum who originally encouraged me to write the "Habilitationsschrift", and to M. SOSTMANN und C. THELEN who typed the manuscript. Last but not least I want to thank my wife Silke for her encouragement during months when I spent much time with this book.

Jülich, Germany   Ralf Littke

June 1993

CONTENT

1. INTRODUCTION

The major objective of this text is to provide information on characteristics of organic matter in sedimentary rocks and to relate these characteristics to processes relevant in their deposition, diagenesis, and weathering.

Organic matter derived from biological precursors is a constituent of most sedimentary rocks. Its amount in recent and subrecent sediments is known to be controlled by a variety of physical and chemical factors such as transport distance from the site of plant growth to the site of deposition or oxygen content of water at the sediment/water interface. It is generally assumed that the same factors also influenced the concentration of organic matter in ancient sediments, i.e., the deposition of petroleum source rocks. This actualistic principle can, however, not be used for the explanation of all source rocks. For example, shelf areas were probably much greater than at present during and after times of intense transgression such as at the end of the lower Toarcian or at the Cenomanian/Turonian boundary. Another example is the occurrence of widespread coals derived from tropical peats in the Upper Carboniferous which is an effect of a much greater continental lowland area in the tropical belt than at present.

In view of their economic significance, the evaluation and quantification of the most important parameters that govern the deposition of sediments rich in organic matter is a major goal in organic sedimentology (Huc, 1988). Studies on recent and subrecent sediments should mainly emphasize the relationship between environmental controls and the characteristics of the sediments. Typical examples are

1) the correlation of primary productivity in surface water with the organic carbon content of the recent sediments;
2) the correlation of sedimentation rate with the organic carbon content (both described in Müller and Suess, 1979);
3) the correlation of bottom water anoxia with the organic carbon content (Emeis et al., 1991).

The same environmental factors are also correlated with other sedimentary features such as

1) the occurrence of benthic fossils; i.e. the degree of bioturbation or lamination (Tyson and Pearson, 1991);
2) the petrographic composition of organic particles as a measure of marine vs terrigenous sedimentation of organic matter (Stein et al., 1989);
3) the chemical composition of organic matter as a measure for its degree of pre- and syndepositional degradation.

Many of the above relationships were studied in the past decade. Nevertheless, it is still not possible to derive a comprehensive, deterministic numerical model that is able to predict the deposition of organic matter-rich sediments or to calculate where and when these sediments were deposited in the past. The development of such a model, its application, and refinement would be a major step forward in the field of organic sedimentology.

On the other hand, knowledge on the relationships between environmental parameters and sedimentary features in recent or young sediments can be used to interpret sedimentary features of ancient sedimentary rocks. Several studies aiming in this direction are gathered in the books on "Modern and Ancient Continental Shelf Anoxia" (Tyson and Pearson, 1991) and "Deposition of Organic Facies" (Huc, 1990). These studies may ultimately lead to an understanding of facies variations in organic matter-rich sediments, e.g., of lateral variations of the petroleum generation potential of source rocks. Source rock qualification can become a more powerful tool in hydrocarbon exploration than today.

As important as the deposition of organic matter-rich sediments is their fate during diagenesis. This term is here used to describe post-depositional changes in general including the catagenetic (oil generation) and metagenetic (gas generation) changes. Organic matter is biologically and chemically degraded before, during and after deposition. Products of early diagenetic alteration are mainly carbon dioxide and methane. At greater depth and at higher temperatures, organic matter is gradually converted into a solid residue and liquid and volatile products such as oil and gas (Tissot and Welte, 1984).

To quantify the degree of organic matter conversion was one of the major topics of organic geochemistry and organic petrology in the past decades. In this context, numerous maturity parameters were developed. The most widely used parameter, vitrinite reflectance, is not only used to predict the degree of diagenetic organic matter alteration, but became also the standard calibration parameter for numerical simulations of temperature histories of sedimentary basins (Waples, 1980; Waples et al., 1992 a,b). Although the application of these simulations often suffers from the lack of important input data, much progress has been achieved in this area in recent years; especially the time- and temperature-dependence of vitrinite reflectance is currently a matter of intense scientific discussion (Barker and Pawlewicz, 1986; Larter, 1989; Sweeney and Burnham, 1990). Nevertheless, most currently available data sets only exist for the "oil window" in which relatively low temperatures prevail (Tissot and Welte, 1984). More calibration work is required, when organic maturity parameters will be used in modelling geothermal regimes of sediments that experienced high temperatures (>150°C).

Another important part of the diagenetic history of organic matter-rich sediments is the migration of oil and gas. Most petroleum source rocks are fine-grained and characterized by low permeabilities; nevertheless oil molecules which are partly larger than pore throats (Tissot and Welte, 1984: 306) are able to escape from these source rocks. Basic questions concerning this "primary migration" include 1) by which mechanism and along which pathways is oil driven out of organic matter-rich rocks; 2) how much generated oil and gas can migrate, how much remains in a source rock and 3) what are the differences in migration mechanisms and efficiencies between different types of source rocks, e.g. between siliciclastic rocks, carbonates, and coals. No clear answers seem to exist for these questions, but the ongoing research has provided important hints. In this text, only the migration of oil and gas from coals which are the least permeable and probably most controversially discussed of all source rocks will be addressed in detail.

Finally, for the evaluation of the petroleum generation potential or the maturity of organic matter-rich rocks, the degree of weathering is of interest. Often, weathered outcrop samples are the only material that can be analysed, but geochemical features may be vastly different than in buried samples of similar lithology and maturity (Baker, 1962; Clayton and Swetland, 1978; Leythaeuser, 1973). Accordingly, in order to make predictions on samples at depth based on data on outcrop samples, the effects of weathering on organic matter-rich sediments have to be studied and quantified.

This study on deposition, diagenesis, and weathering of organic matter-rich sediments is a summary of results obtained during seven years of research work at the Institute of Petroleum and Organic Geochemistry (ICG-4) in Jülich, Germany. During this period, organic matter in many sediments representing different depositional environments and diagenetic pathways was studied (Table 1). In order to describe these investigations in the form of a homogeneous manuscript, part of the results will be incorporated into two overview sections on the topics of "deposition of organic matter-rich sediments" and "petroleum generation" (Chapters 3 and 5). Both overview sections are followed by more specific sections on deposition of a "classic" petroleum source rock (Posidonia Shale) and on "microscopic and sedimentologic evidence for the generation and migration of petroleum" (Chapters 4 and 6). Chapter 7 on "migration of oil and gas in coals" is an overview section with strong emphazis on studies on the coal-bearing Carboniferous from the Ruhr area, western Germany. Aspects of weathering are treated in a much shorter way at the end of this book.

Furthermore, the presented results are restricted to bulk geochemical, organic petrological and sedimentological data. Data on molecular geochemistry are - with few exceptions - neglected, although they provide important additional information.

Lithology	Environment	Stratigr. Interval	Location
Alumshale	Marine	Cambrian	Scandinavia
Coals	Fluvio-Deltaic	Upper Carboniferous	Northwest Germany
Siltstones	Fluvio-Deltaic	Upper Carboniferous	Northwest Germany
Sandstones	Fluvio-Deltaic	Upper Carboniferous	Northwest Germany
Coals	Fluvio-Deltaic	Carboniferous-Permian	Madagascar
Siltstones	Fluvio-Deltaic	Carboniferous-Permian	Madagascar
Black Shales	Marine	Upper Carboniferous	North America
Sandstones	Continental	Permian	Northern Germany
Posidonia Shale	Marine	Toarcian (Jurassic)	Northern Germany
Posidonia Shale	Marine	Toarcian (Jurassic)	Southern Germany .
Asphaltic Limestones	Marine	Malm (Jurassic)	Northern Germany
Kimmeridge Clay	Marine	Malm (Jurassic)	North Sea
Siltstones, Black Shales	Marine	Triassic-Cretaceous	Offshore NW-Australia (ODP Legs 122+123)
Black Shales	Marine	Cenom.-Turon. (Cretaceous)	Morocco and Galicia Margin (ODP Leg 103)
Black Shales	Lacustrine	Eocene (Tertiary)	Southern Germany (Messel)
Black Shales	Lacustrine	Miocene (Tertiary)	Southern Germany (Nördlinger Ries)
Oozes and chalks	Marine	Cretaceous-Quaternary	Central Indian Ocean (ODP Leg 121)
Oozes and clays	Marine	Neogene	Offshore NW-Africa (ODP Leg 108)
Clays and silts	Marine	Neogene	Baffin Bay (ODP Leg 105)
Clays and oozes	Marine	Neogene	Offshore Peru (ODP Leg 112)
Clays and oozes	Marine	Neogene	Offshore Oman (ODP Leg 117)
Clays and silts	Marine	Neogene	Japan-See (ODP Leg 128)

Table 1:   Lithology, depositional environment, stratigraphic interval, and location of sample series treated in this study.

2.   METHODS TO STUDY SEDIMENTARY ORGANIC MATTER

2.1   Overview

The amount of organic matter in rocks is generally calculated after measuring the weight percentage of organic carbon (C_{org}, TOC) on rock powders or on extracted rock powders (Fig. 1). The carbon percentage of sedimentary organic matter usually varies between 50% (e.g., in lignites and peats) and 90%

(e.g., in anthracites). Organic hydrogen, sulphur, nitrogen, and oxygen are generally not measured. The total amount of organic matter can also be estimated by point counting of organic particles in polished sections.

Sedimentary organic matter is subdivided into <u>kerogen</u> and <u>bitumen</u> (Durand, 1980) which are defined as the parts of the organic matter which are insoluble and soluble, respectively, in an organic solvent. This subdivision is widely used in organic geochemistry, although it is incomplete, because the solvent and further details of the extraction are not specified (Tissot and Welte, 1984:131).

A general scheme of the procedures used in characterizing kerogen and bitumen is shown in Fig. 1. In most rocks, kerogen forms more than 90% of the total organic matter. By standard procedures it cannot be chemically characterized in the same detail as bitumen, i.e., on a molecular level. Kerogen is analyzed by elemental analysis, by microscopic methods, by pyrolysis, by spectroscopic methods and by chemical degradation. The latter two techniques were recently reviewed by Rullkötter and Michaelis (1990) in much detail. Here, only a short introduction to microscopic and pyrolytic techniques will be given, because data obtained by these methods are often used in the following text.

Bitumen is obtained by solvent extraction of rock powders and is generally separated into four compound groups by medium pressure liquid chromatography (MPLC; Radke et al., 1980): Saturated hydrocarbons, aromatic hydrocarbons, heterocompounds, and a residue. These classes can be characterized on a molecular level by gas chromatography and gas chromatography coupled with mass spectrometry (Fig. 1). More information on the specific methods used to characterize the sediments treated here is found in Rullkötter et al. (1988a) and Littke et al. (1990).

2.2   Microscopy (Organic Petrology)

Microscopy of organic particles (=macerals) in sediments is a modern scientific discipline derived from coal petrology and palynology (Teichmüller, 1986). The approach derived from coal petrology requires that blocks of sediments, preferentially oriented perpendicular or parallel to the bedding plane, or rock pieces (cuttings) or kerogen concentrates are embedded in a resin, ground flat, and polished. The polished blocks are studied in incident light at high magnification (200-1000x). The palynological approach is based on the preparation of strew slides from kerogen concentrates which are then investigated in transmitted light, usually at lower magnification. In the following, only the incident light technique will be presented and used.

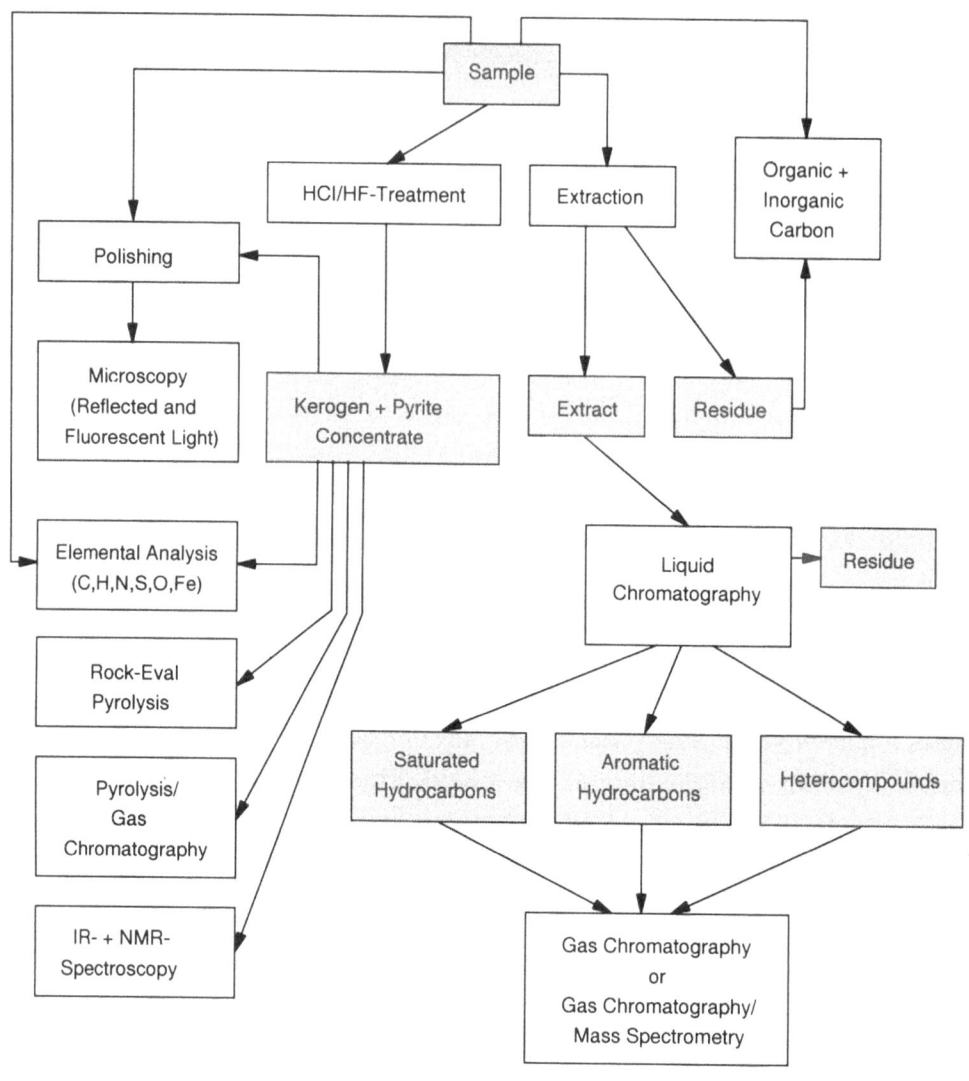

Fig. 1:   Flow chart for analysis of organic matter in sedimentary rocks.

In incident light three maceral groups (inertinite, vitrinite, liptinite) are distinguished based on their reflectivity. Each of these groups consists of various macerals which are described in detail by Stach et al. (1982). At low levels of maturation, inertinite reflectance is greater than vitrinite reflectance and much greater than liptinite reflectance. With increasing maturation, reflectivities of all macerals rise and become more and more similar. In correspondence to these optical changes, the chemical properties of the three maceral groups differ greatly at low levels of diagenetic alteration and become similar with increasing maturation. For example, atomic hydrogen/carbon ratios of liptinites are higher than those of vitrinites and inertinites at immature stages (Fig. 2A). With increasing maturation, all maceral groups preferentially lose hydrogen and oxygen and evolve towards a carbon-rich product (Fig. 2A; van Krevelen, 1961). For the same three maceral groups, hydrogen index and oxygen index evolution are shown in Fig. 2B (see next section).

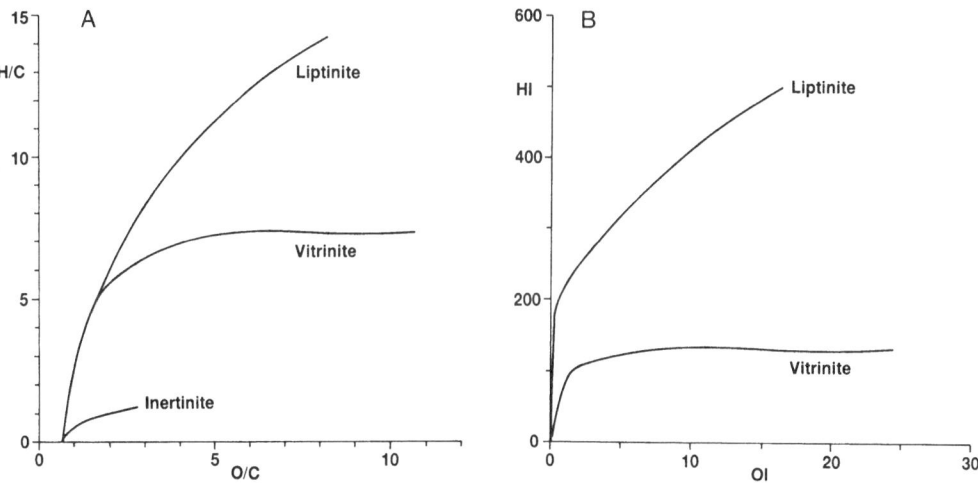

Fig. 2:   A: Hydrogen/carbon and oxygen/carbon atomic ratios of the three maceral groups liptinite, vitrinite, and inertinite (after van Krevelen, 1961). H/C and O/C ratios decrease with increasing maturation. B: Hydrogen index (HI; mg hydrocarbon-equivalents/g C_{org}) and oxygen index (OI; mg CO_2/g C_{org}) values of vitrinites and liptinites. HI and OI values were calculated from H/C and O/C ratios (Fig. 2A) using the H/C-HI and O/C-OI correlation of Espitalié et al. (1977) with a modification for OI-values below 3.

Generally, vitrinites and inertinites are distinctly visible in incident white light and can be easily distinguished from minerals. In contrast, liptinites are dark in reflected light and hardly visible if surrounded by dark minerals such as clays. Fortunately, liptinites fluoresce intensely if irradiated by ultraviolet or violet light. Accordingly maceral countings as used in this study were performed in reflected white light to count vitrinite and inertinite (and pyrite) and in a fluorescent mode to count liptinite. One notable exception is bituminite that belongs to the liptinite group, but which due to its low fluorescence intensity was generally counted in reflected white light. A comparison of total counted macerals (expressed as volume of whole rock) and C_{org}-values reveals the amount of the total organic matter which is visible by optical microscopy or invisible, i.e., submicroscopically small (Fig. 3). Theoretical values for volume percentages of organic matter can be calculated by converting weight -% C_{org} into weight -% OM (organic matter).

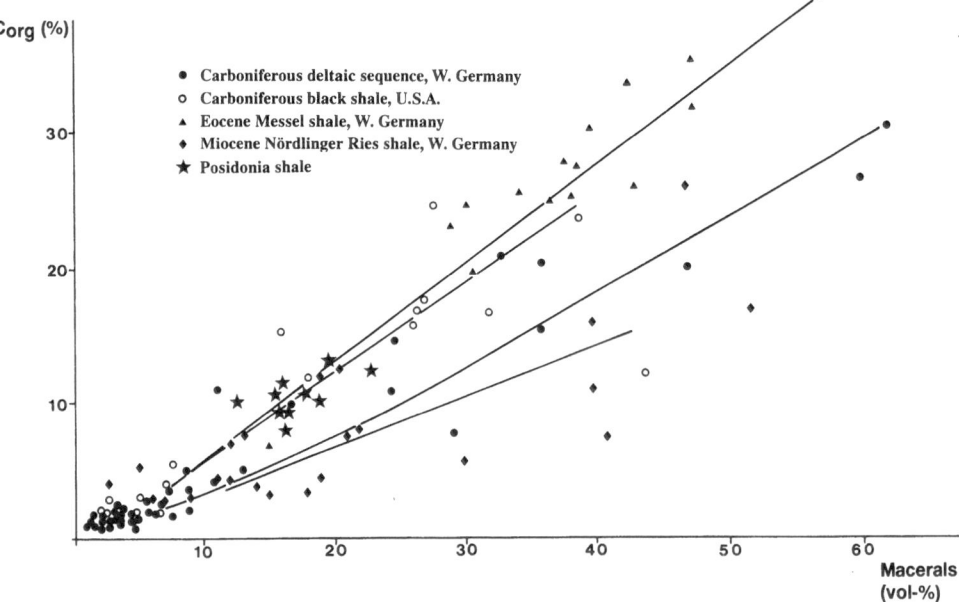

Fig. 3: Volume percentage of total macerals counted in incident white light and in a fluorescence mode plotted versus organic carbon weight percentages. In most sedimentary rocks, the bulk (>80%) of the organic matter is visible as microscopic particles (= macerals; see text for explanation).

$$OM\ (\%) = C_{org}\ (\%)\frac{100}{\%C}$$

In this equation, %C is the percentage of carbon in the organic matter. Furthermore weight -% OM has to be converted into volume -% OM according to

$$OM\ (vol\ \text{-}\%) = OM\ (\%)\frac{\varphi\ (rock)}{\varphi\ (OM)}$$

In this equation, φ is the density of the rock. The ratio φ (rock)/φ (OM) can be expected to vary between 1.6 and 2.0 in most cases. Smaller values have to be assumed for rocks which are extremely rich in organic matter.

As an example, some clastic, deltaic Carboniferous sediments from the Ruhr area (see Fig. 3) are roughly characterized by φ(rock)=2.5 (g/cm^3 or t/m^3), φ(OM)=1.3, %C=70. C_{org}-values are at 3%. The calculated volume percentage of organic matter is 8.4% for these rocks. These theoretical values can then be compared to counted volumes of organic particles to determine the percentage of submicroscopical or optically non-discrete organic matter (Fig. 3). This percentage varies considerably for different organic matter-rich sediments. In most sediments, more than 80% of the total organic matter occurs in the form of discrete, optically resolvable particles. The percentage is in some sediments much smaller, in which marine organic matter predominates, especially in young sediments in coastal upwelling areas such as offshore Oman or Peru (< 10%, ten Haven et al., 1990).

The two most widely used optical maturity parameters are vitrinite reflectance and liptinite fluorescence. Vitrinite reflectance was measured in oil immersion for light of about 546 nm wavelength as described by Stach et al. (1982). Fluorescence of macerals was stimulated by irradiation with uv/violet light and spectra were evaluated according to the procedures described by Teichmüller (1982). More information on the microscopic instrumentation is found in Littke et al. (1988).

2.3   Rock-Eval pyrolysis

Besides microscopy, pyrolytic techniques are well suitable for characterizing organic matter. The most widely used method is the Rock-Eval pyrolysis as described by Espitalié et al. (1977). In this device a sample held in an inert atmosphere is first heated to 300°C (Fig. 4). At this temperature, a constant helium flow extracts all mobile organic compounds. The amount of these compounds is measured by a flame ionisation detector (S_1). Subsequently, the sample is further heated at a constant rate of 25°C/min to an end temperature of 550°C (Fig. 4). During this pyrolysis process, hydrocarbons, other organic compounds and CO_2 are generated from the degrading organic matter (kerogen). In a constant helium stream these volatile products are transported out of the oven and split into two fractions. One fraction is immediately led to a flame ionization detector where the amount of organic products is measured (S_2), whereas the other part is kept in a trap and later released to a thermal conductivity detector to measure the total amount of CO_2 (S_3). S_1 and S_2 are converted into milligrams of hydrocarbon-equivalents per gram of rock (mg hc/g rock), S_3 is expressed as milligrams CO_2 per gram of rock. Derived from the S_2 and S_3 values and the organic carbon content are

- the Hydrogen Index value   (HI = mg hc/g C_{org})
- the Oxygen Index value   (OI = mg CO_2/g C_{org}),

which are used to characterize kerogen in a similar way as atomic hydrogen/carbon and oxygen/carbon ratios (Espitalié et al., 1977; see Fig. 2). Additional parameters used to characterize the maturation state of organic matter are the Production Index or Transformation Ratio (PI=$S_1/(S_1+S_2)$) and the temperature of maximum pyrolysis yield T_{max} (°C, see Fig. 4). Both values increase with increasing maturation.

Besides the Rock-Eval technique, pyrolysis-gas chromatography (see Fig. 1) is now widely used to obtain a molecular characterization of the hydrocarbons generated during pyrolysis (Horsfield, 1989). Pyrolysis has become one of the most useful tools in organic geochemistry. It does, however, not mimic the natural petroleum generation process for which the temperature and time regime are completely different. Also, the role of the presence or absence of minerals on pyrolysis results is of great importance (e.g. Peters, 1986). Generally, mixtures of minerals with organic matter produce much lower HI-values, different OI-values, and higher T_{max}-values than pure organic matter. An example of the first effect is shown in Fig. 5 (Littke et al., 1992) for a series of artificial mixtures of specific minerals (90%) and lignite (10%). The last column of this histogram gives the HI-values of the original lignite (100%). The differences between the HI-values are due to the adsorption of generated hydrocarbons on mineral surfaces (mineral matrix effect) or due to reactions between hydrocarbons and breakdown products of minerals which are not stable under the pyrolytic conditions.

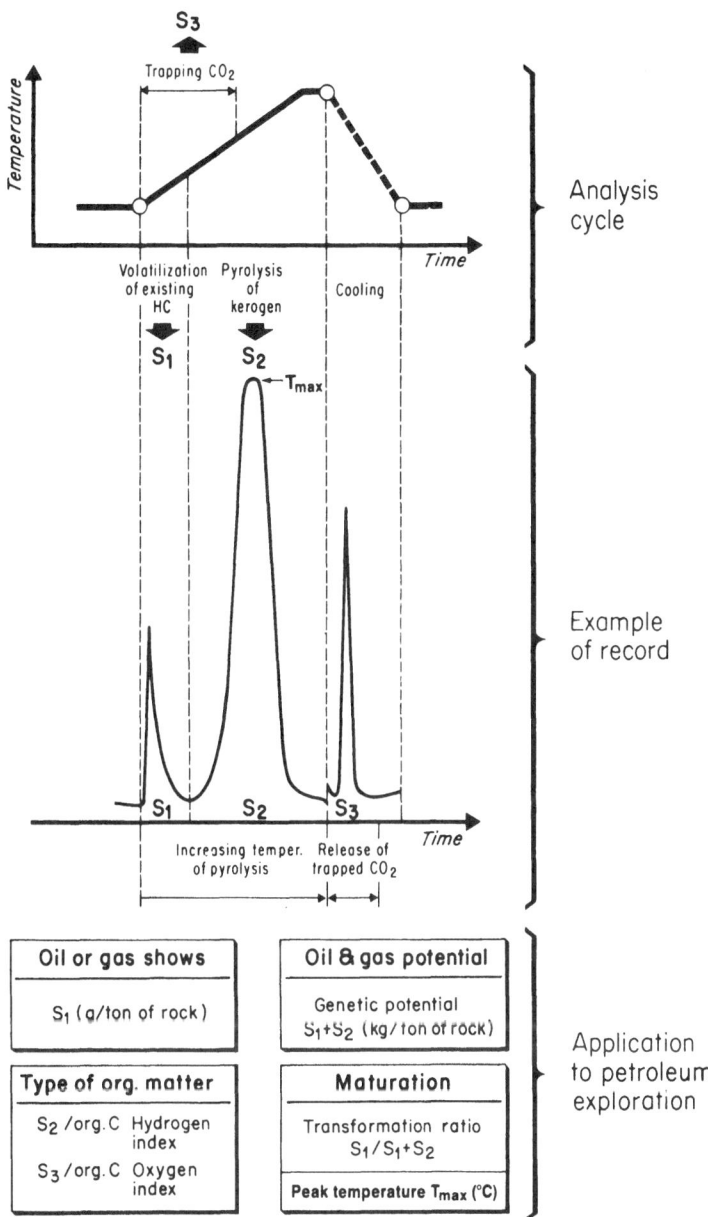

Fig. 4:   Principle of the Rock-Eval pyrolysis (from Tissot and Welte, 1984).

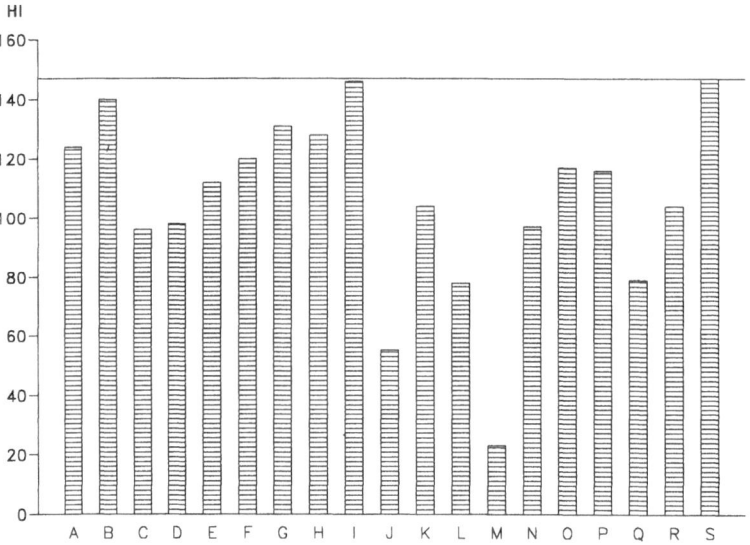

Fig. 5:   Hydrogen index (HI) values (see Fig. 4) of a lignite sample (S; 56% C_{org}) and
                  mixtures of 10 weight% of the same lignite with different minerals (90%).

                  A: calcite, B: dolomite, C: siderite, D: gypsum, E: anhydrite, F: quartz, G: albite, H:
                  microcline, I: illite, J: montmorillonite, K: kaolinite, L: pyrite, M: goethite, N:
                  haematite, O: calcite + illite + quartz + pyrite, P: dolomite + illite + quartz + pyrite,
                  Q: siderite + illite + quartz + pyrite, R: anhydrite + illite + quartz + pyrite. The C_{org}-
                  value of all samples (except S: 56%) is about 5.6%.

3.   DEPOSITION OF ORGANIC MATTER-RICH SEDIMENTS

3.1 Overview

The deposition of sediments rich in organic matter (C_{org} roughly greater than 1%) is usually restricted to

subaquatic sedimentary environments in which organic matter is produced faster than it can be destroyed

(Tourtelot, 1979). In Recent times, those environments generally have in common that they are situated

land-near, i.e. less than a few hundred kilometers away from at least one coastline (Kruijs and Barron,

1990) and that bottom waters are oxygen-depleted if compared to surface waters. Subaerial accumulation

of organic matter is restricted to raised bog peats that are growing in humid climates and which are

considered as important coal precursors (Smith, 1957, 1962). In raised bogs, the water-table is held at the

sediment/air interface and the water is extremely acidic (pH 3-4), thus inhibiting microbial activity and

high rates of remineralisation of plant tissues.

Two contrasting models (Pedersen and Calvert, 1990; Demaison, 1991; Pedersen and Calvert, 1991) are mainly used to explain the deposition of subaquatic, fossil organic matter-rich sediments which will be called "stagnation model" and "productivity model" in the following (Fig. 6; Thunell et al., 1984). The "stagnation model" assumes deposition under a stratified water mass, of which the bottom part is anoxic or suboxic (Table 2; Tyson and Pearson, 1991).

Oxygen (ml/l)	Environments	Biofacies	Physiologic regime
8.0-2.0	Oxic	Aerobic	Normoxic
2.0-0.2	Dysoxic	Dysaerobic	Hypoxic
2.0-1.0	moderate		
1.0-0.5	severe		
0.5-0.2	extreme		
0.2-0.0	Suboxic	Quasi-anaerobic	
0.0 (H_2S)	Anoxic	Anaerobic	Anoxic

Table 2:   Terminology for low oxygen regimes and the resulting biofacies from Tyson and Pearson (1991). The authors suggest that the boundary between suboxic and dysoxic conditions marks the limit for bioturbation (see Savrda and Bottjer, 1991) and nitrate reduction.

Recent examples are the Black Sea and Lake Tanganyika (Huc, 1988a). Essential for the validity of this model is a small rate of water exchange, especially in the vertical direction; high rates would ultimately result in an oxygenation of bottom waters. It is generally assumed that small rates of water exchange are favoured in silled basins (Demaison and Moore, 1980). Limited degradation of organic matter in the absence of oxygen rather than high bioproductivity is the principal base of the "stagnation model". It should, however, be noted that most coastal areas or shelf seas are characterized by higher than average bioproductivity (Huc, 1988a; Kruijs and Barron, 1990), i.e. the model is generally applied to sedimentary rocks for which medium to high bioproductivity rates can be assumed.

The "productivity model" (Fig. 6) is based on high biological productivity in surface waters as recently observed in upwelling areas, e.g., offshore Peru (Thiede and Suess, 1983 a,b; Miller, 1989; Pedersen and Calvert, 1990). Essential for this model are high nutrient supply and sufficient sunlight. It should be noted that the great masses of decaying organic matter also cause oxygen deficiency in the bottom water of these environments (Fig. 6). In contrast to the "stagnation model" strong currents may occur.

It is, however, not entirely proven, whether the same conceptual settings also created organic matter-rich aquatic sediments in the past. For example, it is known that enhanced atmospheric CO_2 concentration leads to enhanced marine bioproductivity (Kruijs and Barron, 1990). Such conditions may have favoured the deposition of organic matter in the past. Also, transgressive times may enhance burial of organic carbon in shelf regions (Wenger and Baker, 1986), whereas times of regression may promote organic matter accumulation in prograding deltaic fans in deep water. For the understanding of organic matter deposition, such time-dependent geological processes are probably of great importance.

In contrast to marine and lacustrine sediments, deltaic and fluvial sediments and coals generally do not contain much organic matter derived form aquatic organisms. However, they often contain sizeable quantities of tissues of higher plants which are less easily destroyed by oxidation and bacterial degradation (Tissot and Welte, 1984). This type of organic material is less hydrogen-rich and less oil-prone than the aquatic type.

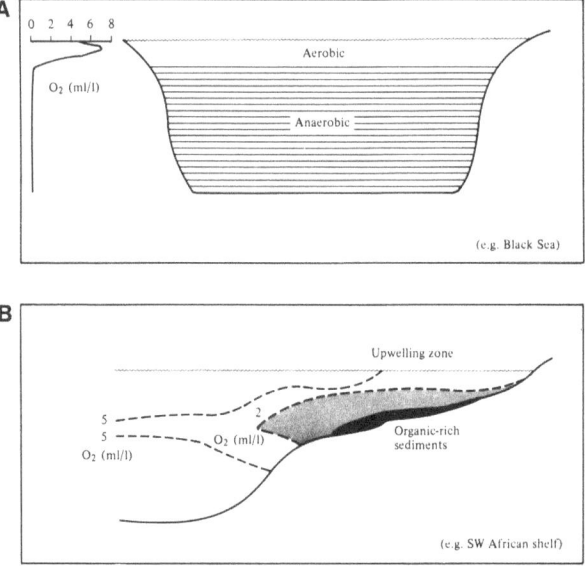

Fig. 6:   Generalised models for the deposition of organic matter-rich sediments. A: Stagnation model for anoxic silled basins, B: Productivity model for upwelling regions (modified after Littke and Welte, 1992 and Thunell et al., 1984).

In summary, balanced optimum conditions between the energy level in a body of water, the supply of nutrients and other factors are required for deposition of organic matter-rich sediments, i.e., of excellent hydrocarbon source rocks. Continent-near, silled basins or vast shelf areas are well-suited for their deposition. It is of special interest that possible reservoir rocks may be expected to occur in these same environments: deltaic or beach sands or reefs are often deposited in the same type of basin just separated from the organic matter-rich sediments by space and time, i.e., by a few kilometers or a few million years. Less oil-prone organic matter may, however, be deposited in a variety of sedimentary environments, e.g., in fluvial and deltaic sediments.

In the following, aspects of the transport of organic particles are discussed and sedimentary features and organic-geochemical and -petrological characteristics of sediments from seven different environments will be presented in the light of environmental controls.

3.2   Transport of organic particles

Most of the biologically produced organic matter is transported after its death by wind or water. Exceptions are plants in swamps and raised bogs and benthic algal and bacterial life forms (microbial mats) that can be incorporated into sediments at the site of growth without transportation. During transportation the degradation of the organic matter starts and the most labile parts of the organic matter are remineralized. This chemical degradation is enhanced by mechanical breakdown and a decrease in particle size causing higher surface/volume ratios. Though the transport of organic particles preserved in sediments occurs either in water or first by wind and later in water, the following discussion is restricted to transport in water.

The process of transportation of organic matter follows the same principle rules as the transport of mineral grains (e.g., von Engelhardt, 1973). The most simple calculation of transport is the determination of the sinking velocity (vs) of an ideal spheric particle according to Stokes' formula

$$vs = \frac{(\varphi_2 - \varphi_1) \cdot g \cdot D^2}{18\,\eta},$$

where φ_2 and φ_1 are the densities ($kg \cdot m^{-3}$) of the particle and water, g is the earth acceleration (gravity; $m \cdot s^{-2}$), D is the particle diameter (m) and η is the dynamic viscosity ($kg \cdot m^{-1} \cdot s^{-1}$). According to the above equation, the sinking velocity of terrigenous organic particles (vitrinite precursors) of 100 µm diameter is about 1.6 mm/s in freshwater. In a 1000 m deep lake such a particle would reach the sedimentary surface within 7 days, if stationary conditions are assumed (Table 3). Terrigenous organoclasts of equal shape and density but only 10 µm in diameter would need almost 2 years to sink to the bottom and algal particles that have a lower density than terrigenous particles travel even longer. It should be noted that real sinking velocities depend on the grain shape. Von Engelhardt (1973: Fig. 3.4) showed that the sinking velocity of quartz spheres of 10 - 100 µm diameter is two orders of magnitude greater than of muscovite plates of equal diameter. Organic particles - in contrast to quartz grains - are rarely spheric and well rounded. The typical shape of terrigenous organic particles in young, open-marine sediments is irregular-cylindric with the longest axis being about twice as long as the shortest axis (Littke et al., 1991c). Therefore lower sinking velocities than calculated in Table 3 have to be assumed for organic particles. Furthermore, compared to freshwater, sinking velocities in sea water are lower due to its greater dynamic viscosity. A more rapid transportation of organic remains is possible, if they are incorporated in large fecal pellets (Tourtelot, 1979), which are known to settle faster than small particles. Degens and Ittekott (1987) strongly advocate the transfer of organic particles in fecal pellets "which are jetted to the sea floor at velocities of about 500 cm/day." Even higher velocities can be assumed following the calculation according to Stokes' formula (Table 3).

Besides theoretical calculations on the transport of particles, sinking rates of aggregated and single organoclasts were discussed in the past (e.g., Riley 1970), but there are only few data on grain size distributions of organic particles in sediments. Littke et al. (1991c) determined the grain sizes of vitrinites and inertinites (i.e., allochthonous, terrigenous organic particles) in marine sediments in the central Indian Ocean and found that there are virtually no particles larger than 20 μm (Fig. 7). Particles of up to 30 μm in diameter were only observed in some pre-Maestrichtian sediments (Site 755) deposited in a palaeogeographic position close to the Cretaceous Kerguelen-Heard plateau, i.e., close to a major island. Such small grain sizes seem to be typical for open marine sediments, if no turbidite transport (see Degens et al., 1986) influenced deposition. In contrast, shallow marine or deep water, continent-near deposits such as the Kimmeridge Clay from the Brae field, North Sea, may contain particles of 50 μm diameter and more (Fig. 7). Even greater grain sizes are found in fluvial-deltaic sediments deposited in humid climatic conditions. For example, upper Carboniferous siltstones and sandstones of the Ruhr area contain abundant plant fragments of several millimetres length (Scheidt, 1988).

Particle Type	Sinking Velocity (m/s)	Time to travel through 1000 m of water	Lateral transport length (m) at current velocity of 10^{-3} m/s and 100 m water depth (m)
Terrigenous Organoclast $\phi = 100\,\mu m$ $\varphi = 1300\;kg/m^3$	$1.6 \cdot 10^{-3}$	7.1 days	62.5
Terrigenous Organoclast $\phi = 10\,\mu m^1$ $\varphi = 1300\;kg/m^3$	$1.6 \cdot 10^{-5}$	1.9 years	$6.25 \cdot 10^3$
Fecal Pellet $\phi = 1\;mm$ $\varphi = 1300\;kg/m^3$	$1.6 \cdot 10^{-1}$	2 hours	$6.25 \cdot 10^{-1}$
Quartz grain $\phi = 10\,\mu m$ $\varphi = 2600\;kg/m^3$	$8.7 \cdot 10^{-5}$	133 days	$1.15 \cdot 10^3$

Table 3:   Sinking velocity, travel time, and lateral transport length for spheric particles in non-turbulent water (see von Engelhardt, 1973). Densities of coal macerals usually vary between 1.1 and 1.7 g/cm^3 (van Krevelen, 1961).

The observations on grain sizes of allochthonous organic particles in different environments indicate that transport of organoclasts leads to sorting of grains according to size and to mechanical grain size reduction with increasing length of transport or transport energy in very much the same way as well established for mineral grains. In general, large transport distances of organoclasts are favoured by their low density compared to mineral grains. This also leads to large residence times of organoclasts in the water.

While large transport distances of terrigenous organoclasts compared to terrigenous mineral matter lead to a relative enrichment of the former, transport sorting will hardly cause deposition of layers rich in terrigenous organic matter at distal sites in the oceans, because more and more autochthonous siliceous and calcareous marine particles will be admixed. It is also not to be expected that liptinite-rich sediments evolve from transport sorting, although liptinites (hydrogen-rich organic particles; see Fig. 2) have a lower density than vitrinites and inertinites. Liptinites are, however, also more liable to degradation and therefore less well preserved if transport lasts long. This is the reason why in most marine sediments deposited under oxic bottom waters terrigenous organic particles (vitrinites and inertinites) predominate over marine particles (alginite).

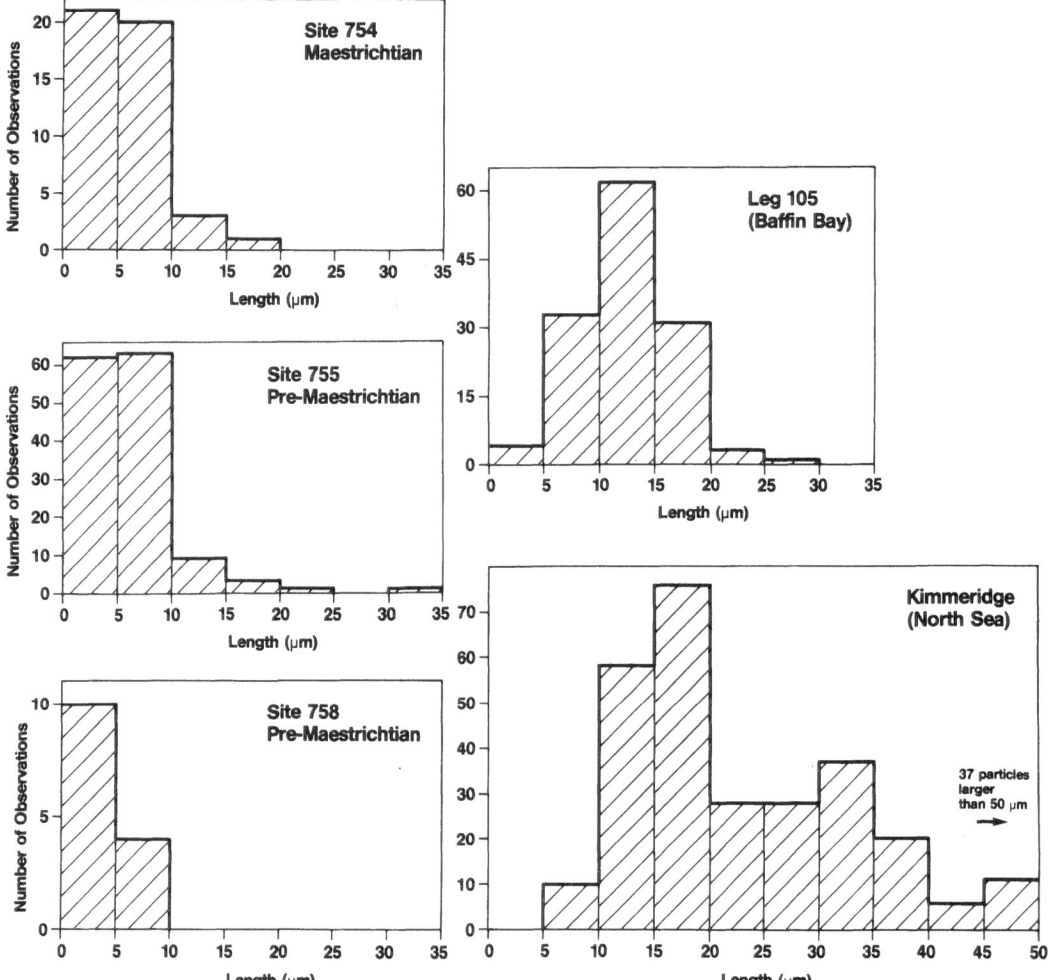

Fig. 7: Histograms of length of terrigenous organic particles (vitrinites and inertinites) at Sites 754 and 755 (Broken Ridge, Indian Ocean), 758 (Ninetyeast Ridge, Indian Ocean), 645 (Baffin Bay, i.e., near land, see Stein et al., 1989) and in the Kimmeridge Clay in the North Sea (Brae Field, see Leythaeuser et al., 1988). The particle size is related to the length of tranport (after Littke et al., 1991c).

Generally, with increasing length of transport, the inertinite/vitrinite ratio increases indicating a greater degree of chemical degradation (Fig. 8). In many coals as well as in many lacustrine, fluvial, deltaic or shallow marine deposits, vitrinite is the most abundant group of terrigenous organoclasts (e.g., Horsfield et al., 1988), but in most open marine deposits inertinite predominates (e.g., Stein et al., 1989). The latter is known to be the product of a more severe degradation of terrigenous organic matter whereas vitrinite derives from a less severe degradation of the same principle precursors (Stach et al., 1982; Styan and Bustin, 1983). It is an open question why at least some small-sized vitrinite is preserved in open ocean sediments derived from slow deposition under well-oxygenated bottom water, i.e., under conditions favorable of a strong degradation of organic matter.

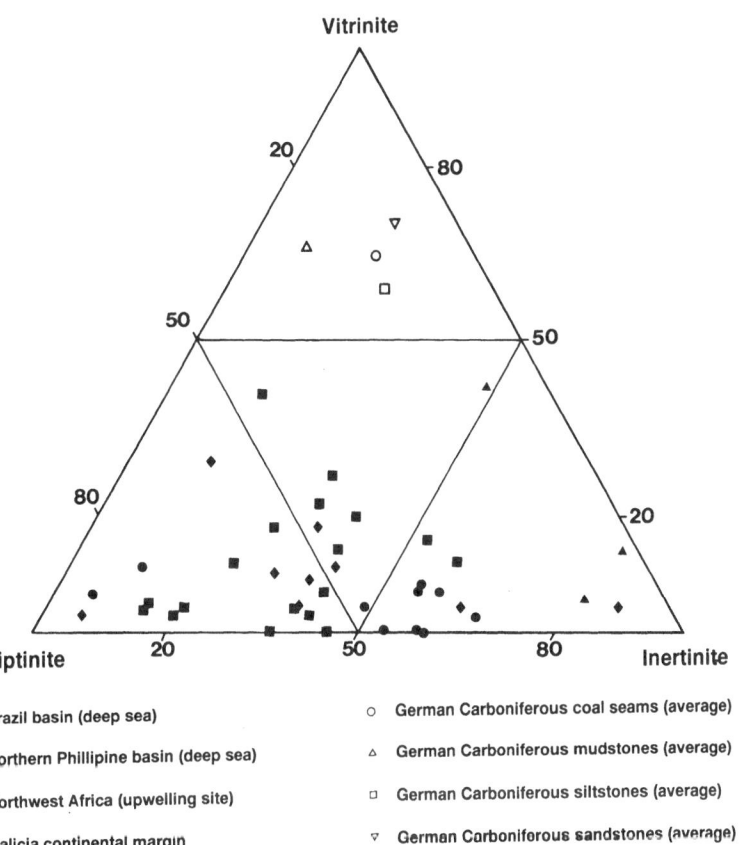

Fig. 8:   Percentages (of total macerals) of vitrinite, inertinite, and liptinite in different environments. Predominance of vitrinite is only to be expected in coals and in fluviodeltaic environments, where transport distances for terrigenous organoclasts are small. Deep sea sediments are often characterized by high inertinite over vitrinite ratios. Data after Mukhopadhyay et al. (1983), Rullkötter et al. (1980), Stein et al. (1988; 1989), and Scheidt and Littke (1989).

3.3   Deep, Marine Silled Basins

Demaison and Moore (1980) identified anoxic silled basins as one major setting for marine oil source rocks and discuss as typical contemporary example the Black Sea. In anoxic silled basins the major control on deposition of organic carbon-rich sediments is anoxicity of bottom water and pore water leading to high ratios of preserved over produced organic carbon (Bralower and Thierstein, 1987).

One typical feature of silled basins is water stagnation. Usually, the upper part of the water is oxygen-rich whereas the bottom water is anoxic or dysoxic (Table 2) and characterized by low bioproductivity. Generally, surface and bottom waters mix only slowly, i.e., there are no strong currents or waves disturbing the water. Deposition usually takes place below storm wave base which can reach as deep as 200m (Galloway and Hobday, 1983), but is usually much shallower (Tyson and Pearson, 1991). Stagnation is often enhanced by contrasting physical or chemical properties of bottom and surface water such as temperature and salinity. Accordingly, the boundary between surface and bottom water is called thermocline or halocline, respectively.

In case of the Black Sea, marine water derived from the Mediterranean is overlain by less saline water of fluvial origin (Danube, Don, Dnepr). The resultant halocline marks also the boundary between oxic and anoxic water. When the anoxic bottom waters were established about 7000 years ago (see Degens et al., 1978 for revision of age data), maximum organic carbon contents of the sediments increased from 0.7 to 20% (Demaison and Moore, 1980). Most sediments in the 40cm thick, laminated black shale layer deposited between 7000 and 3000 years before present contain between 2 and 7% organic carbon (Glenn and Arthur, 1984). Sediments deposited after 3000 years before present are laminated coccolith oozes and contain only 1-5% organic carbon (Shimkus and Trimonis, 1974), although the extension of the bottom water anoxic zone is even greater than during the depositon of the laminated black shales (Deuser, 1974; Pedersen and Calvert, 1990). The smaller organic carbon percentage can, in analogy to observations on ancient black shales (see Chapter 4), be tentatively explained as a dilution effect by coccoliths (see Müller and Blaschke, 1969), which are most enriched in the uppermost Black Sea sediments.

In contrast to numerous observations on oxic deep sea deposits in which sedimentation rate and organic carbon percentage are positively correlated (Müller and Suess, 1979; Littke et al., 1991c), a negative correlation exists for Black Sea sediments (Demaison and Moore, 1980) to an extent that "fields of high organic matter content correspond to areas of low primary production and vice versa" (Fig. 9; Huc, 1988a,b; Shimkus and Trimonis, 1974). This type of observation led Stein (1986) to the suggestion that organic matter-rich sedimentary rocks may be assigned to either anoxic or oxic bottom water conditions by plotting sedimentation rate versus organic carbon percentage (Fig. 9). According to this reasoning anoxic deposits are characterized by high organic carbon percentages and low to high sedimentation rates, whereas other deposits only contain high concentrations of organic matter, if sedimentation rates are also high. It should, however, be noted that in the case of the Black Sea as the "classic" silled anoxic basin, sedimentation rates during black shale deposition were also high (about 20-30cm/1000a) due to the deposition of great masses of terrigenous material (Degens et al., 1978). In such a case oxic and anoxic palaeo-bottom water conditions can only be differentiated, if marine and terrigenous sedimentation rates are calculated and if the marine sedimentation rate is plotted versus organic carbon percentages. Also, high resolution stratigraphy is a prerequisite for the adaption of the classification scheme proposed in Fig. 9.

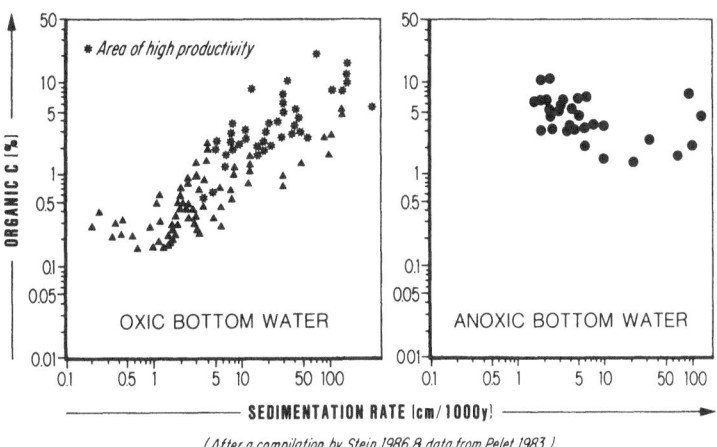

Fig. 9:   Relationship between organic carbon content and sedimentation rate in recent and Quaternary sediments (from Huc, 1988b after Stein, 1986 and Pelet, 1983).

Furthermore, studies by Huc (1988a,b) suggest that in anoxic silled basins, the highest concentration of organic carbon occurs in the deepest parts of the basins leading to a concentric pattern of the organic matter concentration. This is supposed to be an additional criterion for identifying anoxic basins.

3.4   Upwelling Areas

Upwelling areas are regions in which great masses of oceanic surface waters move offshore and are replaced by oceanic deep water. The upwelling process is driven by either atmospheric or ocean forcing mechanisms such as wind, coastal currents, and the Coriolis force (Suess and Thiede, 1983). Upwelling rates are calculated from surface wind stress and the Coriolis force (Kruijs and Barron, 1990) and range from 0 to >20 cm per day. The upwelling deep waters are often enriched in nutrients such as nitrate, phosphate and silicate (Baturin, 1983; Bishop, 1989) thus favouring high primary productivity rates (greater than 180 $gC/m^2/a$; see Berger et al., 1989) in surface waters and oxygen shortage in waters below the photic zone (Fig. 6). Areas of high primary productivity occupy only a very small proportion of the modern oceans (Berger, 1989) and are often related to upwelling. Baturin (1983) estimated that the total area of recent strong upwelling is only 500000 km^2 or about 1 °/oo of the world ocean. Prominent upwelling cells are situated offshore the coasts of Northwest Africa, Southwest Africa, Peru, Northwest America and Oman. The processes influencing sedimentation in upwelling areas are described in great detail in the two volumes of Thiede and Suess (1983) and Suess and Thiede (1983). Typical features of sediments underlying upwelling areas include 1) high organic carbon (2-20%), 2) high biogenic silica (5-70%), 3) elevated phosphorus (0.2->1%) contents, 4) high rates of biogenic sedimentation (up to 0.5 mm/a) and 5) the occurrence of coprogenic material (Baturin, 1983).

Organic carbon accumulation in upwelling areas does not solely depend on primary productivity related to upwelling , but also on the interaction of the upwelling system with the topography of shelf and continental slope (Reimers and Suess, 1983; Kruijs and Barron, 1990). This is supported by Baturin (1983) who suggested that unfavourable conditions for accumulation of biogenic sediments persist where steep shelf platform morphologies, high energy wave climates, strong bottom currents, high rates of terrigenous sedimentation, bioturbation by bottom fauna and repeated transgressive/regressive cycles (erosional events) occur.

A recent example for an upwelling area below which sediments are not particularly enriched in organic carbon is the Northwest African shelf at Site 658 of the Ocean Drilling Program (Table 4, Stein and Littke, 1990), where organic carbon contents generally vary between 1 and 3% and do not exceed 4%. Furthermore, hydrogen index (HI) values from Rock-Eval pyrolysis are relatively low (Stein et al., 1989). This is explained by the occurrence of greater amounts of terrigenous organic matter than at other upwelling sites. In other upwelling areas, the bulk of the organic matter stems from bacterially degraded phytoplankton and terrigenous organic matter contributes only a small proportion to the total organic matter (Summerhayes, 1983; ten Haven et al., 1990). In the case of the Northwest African shelf, the significant contribution of terrigenous organic matter is indicated by petrologic and geochemical data, e.g., by the occurrence of a homologous series of *n*-alkanes with 25 to 31 carbon atoms. Among these, molecules with odd carbon numbers predominate over those with even carbon numbers. Furthermore, the rather low HI-values of the organic matter in these sediments are interpreted to reflect both, high percentages of terrigenous organic matter and a high level of marine organic matter degradation. Undegraded marine organic matter is in general enriched in hydrogen compared to terrigenous organic matter and accordingly displays higher average HI-values.

Area	C_{org} (%)	HI (mg hc/g C_{org})	ARBS ($g \cdot cm^{-2} \cdot yr^{-1}$)	AROC ($g \cdot cm^{-2} \cdot yr^{-1}$)
Site 658 (NW-Africa)	2.0	261	12.0	0.20
Site 723 (Oman margin)	3.1	340	18.3	0.57
Site 679 (Peru margin)	4.0	455	6.5	0.26

Table 4: Average organic carbon percentages, average hydrogen index (HI) values, bulk sediment and organic carbon accumulation rates in sediments on the continental margins offshore Northwest Africa (Pliocene to recent; Stein and Littke, 1990) offshore Oman (Pleistocene to recent; Prell, Niitsuma et al., 1989), and offshore Peru (Quaternary; ten Haven et al., 1990)

Higher organic carbon contents than offshore Northwest Africa are reported from the upwelling area offshore Oman (Tables 4 and 5; Prell, Niitsuma, et al., 1989). Fig. 10 shows an idealized cross section through the continental margin off Oman with the position of six different wells drilled by the Ocean Drilling Program. One of the sites (725) is situated at the very top of the oxygen minimum zone which underlies the surface water. Two sites (723 and 728) are just within the oxygen minimum zone, two sites are on the Owen Ridge and just below the oxygen minimum zone (731 and 722) and one site is outside the upwelling area in the deep sea and influenced by the Indus fan (720). Organic carbon contents and HI-values of the uppermost sediments from these sites are summarized in Table 5 after Prell, Niitsuma et al. (1989) and ten Haven and Rullkötter (1991). Organic carbon percentages and HI-values are greatest in sediments deposited within the oxygen minimum zone, i.e., at water depths which exceed 350 m but are less than 2000 m. Shelf sediments deposited in relatively shallow water like those drilled at Site 725 (311 m) contain little organic matter which is characterized by low HI-values. Most favourable conditions for organic matter deposition occur, where sedimentation takes place in the upper part of the oxygen minimum layer, e.g., at Site 723 at a water depth of 807 m. For these sediments, average HI-values of 298 mg hc/g C_{org} are recorded which are greater than those established for the other sites on the Oman continental margin, but still lower than for many ancient petroleum source rocks.

Site	Depth interval (mbsf)	Strat. Age	C_{org} (%)	HI (mg hc/g C_{org})	ARBS ($g \cdot cm^{-2} \cdot yr^{-1}$)	AROC ($g \cdot cm^{-2} \cdot yr^{-1}$)
725	0 - 38	Holoc.-Pleistoc.	0.56 ± 0.53[1](35)	110 (2)	14.4[3]	0.25
723	0 - 40	Holoc.-Pleistoc.	2.72 ± 1.39[1,2](41)	298 ± 94 (26)	18.3	0.57
728	0 - 65	Holoc.-Pleistoc.	1.60 ± 0.59[2](11)	219 ± 100 (11)	4.0	0.06
731	0 - 50	Holoc.-Pleistoc.	0.65 ± 0.37[1](8)	141 ± 53 (5)	3.8	0.02
722	0 - 53	Holoc.-Pleistoc.	0.97 ± 0.32[1](8)	243 ± 41 (6)	3.2	0.03
720	0 - 48	Holoc.-Pleistoc.	0.56 ± 0.54[1](13)	56 ± 34 (10)	23.1[4]	0.14

Table 5:   Depth intervals (meters below sea floor), stratigraphic age, average organic carbon contents, average hydrogen index (HI) values, accumulation rates for bulk sediment (ARBS) and accumulation rates for organic carbon (AROC) for the uppermost intervals drilled at ODP (Ocean Drilling Program) Sites offshore Oman (see Fig. 10). Note that average C_{org} and HI-values are tabulated for the announced depth interval only, whereas accumulation rates are calculated for the entire Holocene-Pleistocene section. Data are from Prell, Niitsuma, et al. (1989) and ten Haven and Rullkötter (1991).

1   C_{org} was measured by a coulometric method.
2   C_{org} was measured by the Rock-Eval instrument and subsequently corrected (TOCb; see Prell, Niitsuma, et al., 1989).
3   ARBS and AROC were calculated for the entire Holocene-Pleistocene section (163m) with the assumption of an age of 1.35 Million years for the oldest strata (Spaulding, 1991).
4   Calculated for upper 200m, corresponding to the last 1 Million years.

Favourable conditions for the deposition of organic carbon-rich sediments also occur where the sediment/water interface is within the lower part of the oxygen minimum zone or slightly deeper as at Site 722 (about 2000 m). Sediments deposited in the deep sea as at Site 720 (about 4000 m water depth) contain lower average organic carbon contents and show low HI-values. The occurrence of mass flow or turbidity current sediments at these deep locations causes few exceptionally high organic carbon values and is responsible for the high standard deviation recorded in Table 5.

The upwelling system offshore Peru will serve in the following as an example for the complexity of sedimentation patterns in these depositional environments. In upwelling areas, the distribution of organic carbon-rich deposits is not only controlled by primary productivity, but also by current activity and basin morphology (Reimers and Suess, 1983). This is evident from Fig. 11, which compares primary productivity rates in surface water and organic carbon percentages of the uppermost sediments for the Peruvian coast. In general, bioproductivity is most enhanced and rather homogeneously distributed along the coastline and becomes successively smaller towards the open ocean. Organic carbon percentages are exceptionally high in two small areas between 11 and 14°S, but lower towards the north, although upwelling intensity and bioproductivity are in the same range. Reimers and Suess (1983) explain this difference by shallower water depths between 7 and 10°S, which promote continuous reworking by bottom currents and inhibit sedimentation. Maximum bulk sediment and organic carbon accumulation rates occur at about 15°S, in an area where organic carbon percentages of sediments are not exceptionally high. This is explained by Reimers and Suess (1983) as an effect of high fluvial, terrigenous input and high bioproductivity in the south. In the regions with highest organic carbon percentages further north, accumulation rates for both bulk sediment and organic carbon are lower due to the low local fluvial discharge (see Wefer et al., 1990).

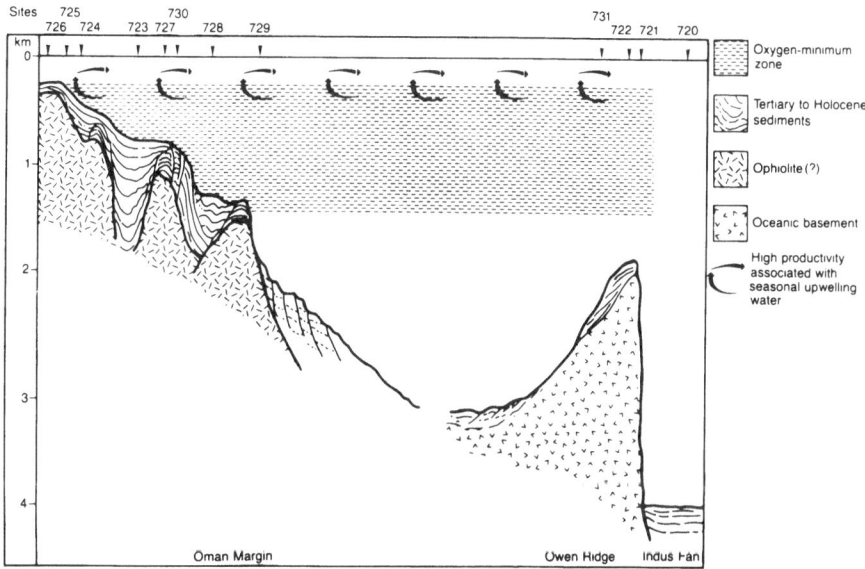

Fig. 10:   Idealized transect of the continental margin off Oman showing major geological and oceanographic features and locations of sites discussed in text (vertical scale exaggerated; from Spaulding, 1991).

Table 6 compares C_{org} and HI-values as well as accumulation rates of a site at about 11°S (area of high C_{org}-contents) and of a site at about 14°S (area of high accumulation rates). Offshore Peru, high C_{org}-

percentages coincide with high HI-values, whereas high accumulation rates do not coincide with high HI-values.

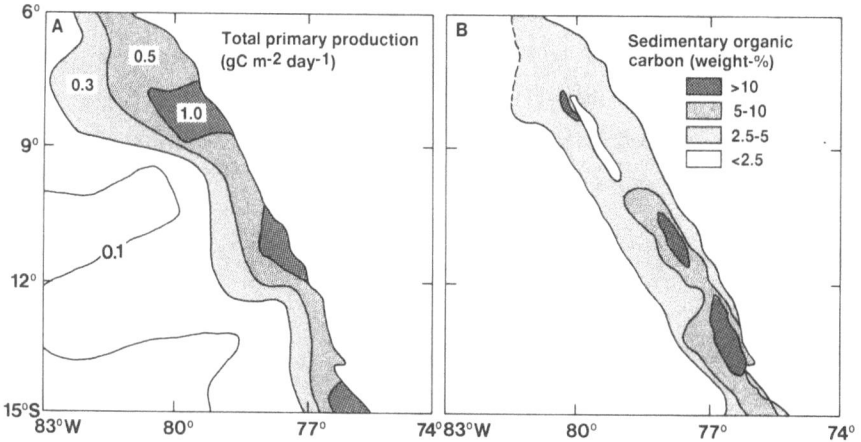

Fig. 11:   Total primary production (A) and sedimentary organic carbon in surface sediments (B) in the upwelling area off Peru (redrawn after Reimers and Suess, 1983).

Comparisons between the occurrences of upwelling on the one hand and high primary productivity on the other reveal that

-       upwelling is the most common but not the only cause for high bioproductivity rates; another source of nutrients is river discharge (van der Zwaan and Jorissen, 1991);

-       many upwelling areas (82%; Kruijs and Barron, 1990) are not associated with high productivity rates, especially in the open ocean. Only along coastlines exists a good positive correlation between upwelling and high productivity (Kruijs and Barron, 1990), possibly due to the favourable topography of the sea floor and/or the additional nutrient supply from land.

According to Summerhayes (1983), "massive, short-period production rather than steady state high productivity controls the accumulation of organic matter in upwelling regimes, by overwhelming the system's ability to recycle organic matter as food in the water column." Other factors besides upwelling and primary productivity also greatly influence organic carbon accumulation and organic carbon percentages in sediments. Interaction of high rates of primary productivity in surface waters and low rates of turbulence in bottom waters and not one or the other is probably the prerequisite for source rock deposition as shown by Reimers and Suess (1983) for the Peruvian continental margin.

Site	Water Depth (m)	Depth Interval (mbsf)	Age (Myr)	C_{org} (%)	HI (mg hc/g C_{org})	ARBS (g·cm^{-2}·yr^{-1})	AROC (g·cm^{-2}·yr^{-1})
680	260	0-41	0-0.62	5.34 ± 2.57(47)	506 ± 177(46)	3.7	0.20
686	450	0-160	0-1.41	2.32 ± 1.36(8)	334 ± 78(8)	19.6	0.45

Table 6:   Depth intervals (meters below sea floor), age intervals (million years), average organic carbon contents, average hydrogen index (HI) values, accumulation rates for bulk sediment (ARBS) and accumulation rates for organic carbon (AROC) for the youngest sediments in Holes 680B and 686B, offshore Peru. Accumulation rates are corrected for the hiatus between 0.62 and 1.37 Myr at Site 686 (Wefer et al., 1990). C_{org} and HI-data are from Suess, von Huene, et al. (1988; Site 686) and Emeis and Morse (1990; Site 680), age information is from Wefer et al. (1990) and sediment densities are from Suess, von Huene, et al. (1988). C_{org} was measured with the Rock-Eval method.

3.5   Anoxic Continental Shelves

Black shales which are deposited in shallow marine environments are common in the Palaeozoic and Mesozoic ancient geological record, but there is no well documented modern analogue (Hallam, 1981:90). Late Carboniferous black shales of Europe and North America are presented here (Table 1), because they are regarded as well-studied examples of these organic matter-rich continental shelf deposits (Heckel, 1991). They clearly represent transgressive phases within the Carboniferous sedimentary sequence in which they are sandwiched within fluvial and deltaic clastic sediments and coals. As these thin (usually 1 m and less) black shales lack any benthic fossils and any evidence of bioturbation (O'Brien, 1987), it is assumed that they were deposited under anoxic conditions (see Table 2).

A geochemical profile through one of these Pennsylvanian black shales is shown in Fig. 12. Organic carbon contents are high (17%) at the base of the black shale and decrease towards the top. HI-values are also greatest at the base (320 mg hc/g C_{org}), but not particularly high, if compared to other immature black shales. Maceral analysis revealed that the organic particles are a mixture of fluorescing alginite probably derived from marine phytoplankton and of brightly reflecting, weakly or non-fluorescing particles (Plate 1, A and B). The presence of the latter material is expected to cause the relatively low HI-values of these black shales. Low HI-values are generally typical for terrigenous organic matter and the "bright particles" were initially proposed to be derived from peat clasts (Wenger and Baker, 1986). Molecular geochemical data and organic petrographic observations (such as the lack of phenols in pyrolysis products and the lack of trimacerites) do, however, not support an origin as terrigenous organic clasts; an explanation as hydrogen-poor humic precipitate is more realistic. The precipitation of humic acids is favoured in environments where water mixing, e.g., mixing of saline water with freshwater, occurs (Swanson and Palacas, 1965; Lyons et al., 1984). Such a mixing is certainly not unprobable for the time of Pennsylvanian black shale deposition.

The organic richness of many thin shallow marine sedimentary rocks which occur in coal-bearing fluviodeltaic strata suggests a common controlling mechanism and asks for a depositional model. Certainly, the two most widely used actualistic models (Fig. 6) do not fit to these deposits, e.g., water depth was certainly much less than in the case of the silled Black Sea (2000 m) and upwelling sediments are different in terms of organic richness, thickness, and other petrographic features (Chapter 3.4). To explain the occurrence of Pennsylvanian black shales of the Northamerican midcontinent, Wenger and Baker (1986) suggested that transgressive-regressive cycles are the major control of the accumulation of these black shale sequences. Rapid transgression of epicontinental seas over laterally adjacent widespread peat swamps and emergent delta-plain surfaces, formed during the previous regressive cycle, resulted in leaching of the flooded sediments and in an enhanced influx of nutrients and humic materials to the marine environment. The nutrients stimulated high algal productivity and the terrestrially-derived organic material provided an additional sink for oxygen which enhanced anoxia. As transgression proceeded to its maximum, the swamps and delta plain were progressively flooded and diminished in areal extent, the influx of nutrients and humic material decreased, productivity dwindled, intensity of anoxia and preservation of organic matter decreased, and black shale deposition terminated. Subsequent regression led to the reestablishment of delta-plain surfaces and coexisting swamp environments creating the conditions appropriate for the deposition of a succeeding black shale during the next cycle of transgression (cf. Wenger and Baker, 1986). This evolution is summarized in Fig. 13 and is in accordance with the geochemical data (Fig. 12) which suggest that black shale deposition starts more abruptly than it ends. Additional evidence for the importance of eutrophication of shallow seas by riverine inflow of nutrients was recently published by van der Zwaan and Jorissen (1991) who concluded that "the chance of anoxia is highest during periods of high sea level, leading to large shelf areas".

29

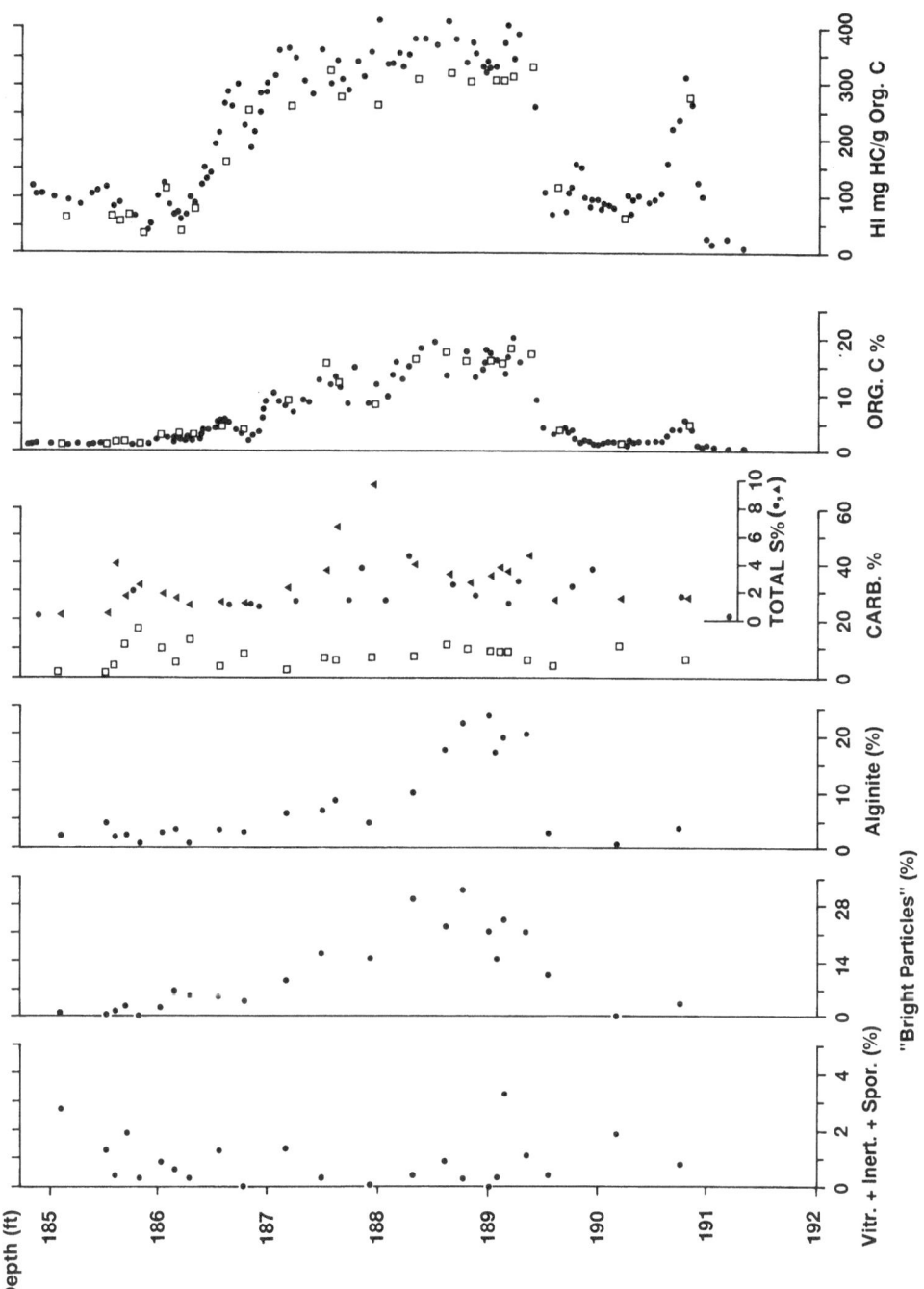

Fig. 12:　Geochemical and petrological profile through the Little Osage Shale sequence in the Kelly well of the Upper Carboniferous of Oklahoma, U.S.A. ($R_r = 0.54\%$). The figure was redrawn from a draft developed in cooperation with D.R. Baker, Houston. In the case of sulphur (S), organic carbon (org. C) and HI-values, additional data from an earlier study (Wenger and Baker, 1986) were also plotted (circles).

30

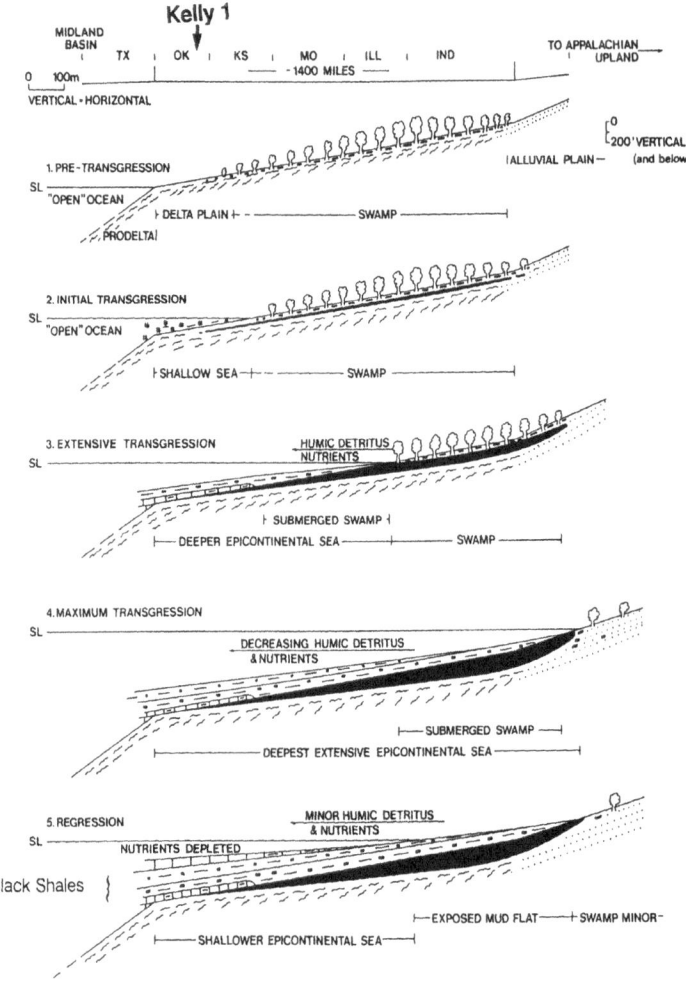

Fig. 13:   Depositional model for transgressive Upper Carboniferous black shales. Inflow from
                 distal, flooded land areas and the leaching of the former land-surface probably
                 provided the nutrients for enhanced algal productivity (see Wenger and Baker, 1986).
                 Humic detritus provided additional (but minor) organic matter. The figure was
                 developed in cooperation with D.R. Baker, Houston.

3.6   Progradational Submarine Fans

Progradational submarine fans are sedimentary sequences which accumulate in deep marine water close to a continent and are usually fed by river discharge. Recent examples are the Mississippi fan, the Indus fan, and the Bengal fan.

C_{org}- and carbonate-concentration profiles through the distal Bengal fan on the northern Ninetyeast Ridge in the Indian Ocean (Littke et al., 1991c; Weissel, Peirce, et al., 1991; Site 758 of the Ocean Drilling Program) demonstrate the evolution from a purely pelagic towards a mixed marine/continental sedimentation (Fig. 14). The Palaeocene and Oligocene pelagic deposits are almost free of organic matter (C_{org} less that 0.2%) as typical for open ocean deposits (Degens and Mopper, 1976) and consist of about 80% carbonate derived from planktonic and benthic organisms. With decreasing depth and increasing progradation of the Bengal fan, there is a decrease in carbonate content and an increase of C_{org}-values. This evolution is enhanced by the ongoing rise of the Himalayan mountain range, causing an increasing sediment discharge by the Ganges and Brahmaputra rivers into the Indian Ocean. The deposited organic matter is mainly of terrigenous origin and hydrogen-poor (Table 7), as in the entire central Indian Ocean (Ninetyeast Ridge and Broken Ridge). In this area, C_{org}-values and organic carbon accumulation rates are positively correlated with sediment accumulation rates (Littke et al., 1991c), as previously established for other oxygenated open ocean sediments by Müller and Suess (1979). This positive correlation is generally interpreted to be an effect of a higher organic matter preservation rate due to a more rapid burial. Müller and Suess (1979) concluded that "as a first approximation, one can expect a doubling of sedimentary organic carbon contents with each 10-fold increase in sedimentation rate". It should be noted that such a positive correlation between organic carbon and sediment accumulation does not exist in all deep marine sediments, for example, it does not exist at Site 721 on Owen Ridge (Fig. 10; Prell, Niitsuma et al., 1989). For the prograding Bengal fan, a further increase in C_{org}-percentages can be predicted for the future.

Compared to the distal Bengal fan, C_{org}-values are slightly higher (0.4-1.1%) in samples from the lower and middle Mississippi fan recovered during DSDP Leg 96 in the Gulf of Mexico (Marzi and Rullkötter, 1986). Like in the Bengal fan, organic matter is mainly of terrigenous origin and HI-values are low (less than 110 mg hc/g C_{org}). The greater HI-values compared to the distal Bengal fan (Table 7) are probably the result of better preservation of terrestrial organic clasts due to shorter transportation distances (about 350 km versus 1800 km between delta and depositional site). Accordingly, the ratio vitrinite/inertinite is high in the Mississippi fan (Marzi and Rullkötter, 1986) and small in the distal Bengal fan (Littke et al., 1991c).

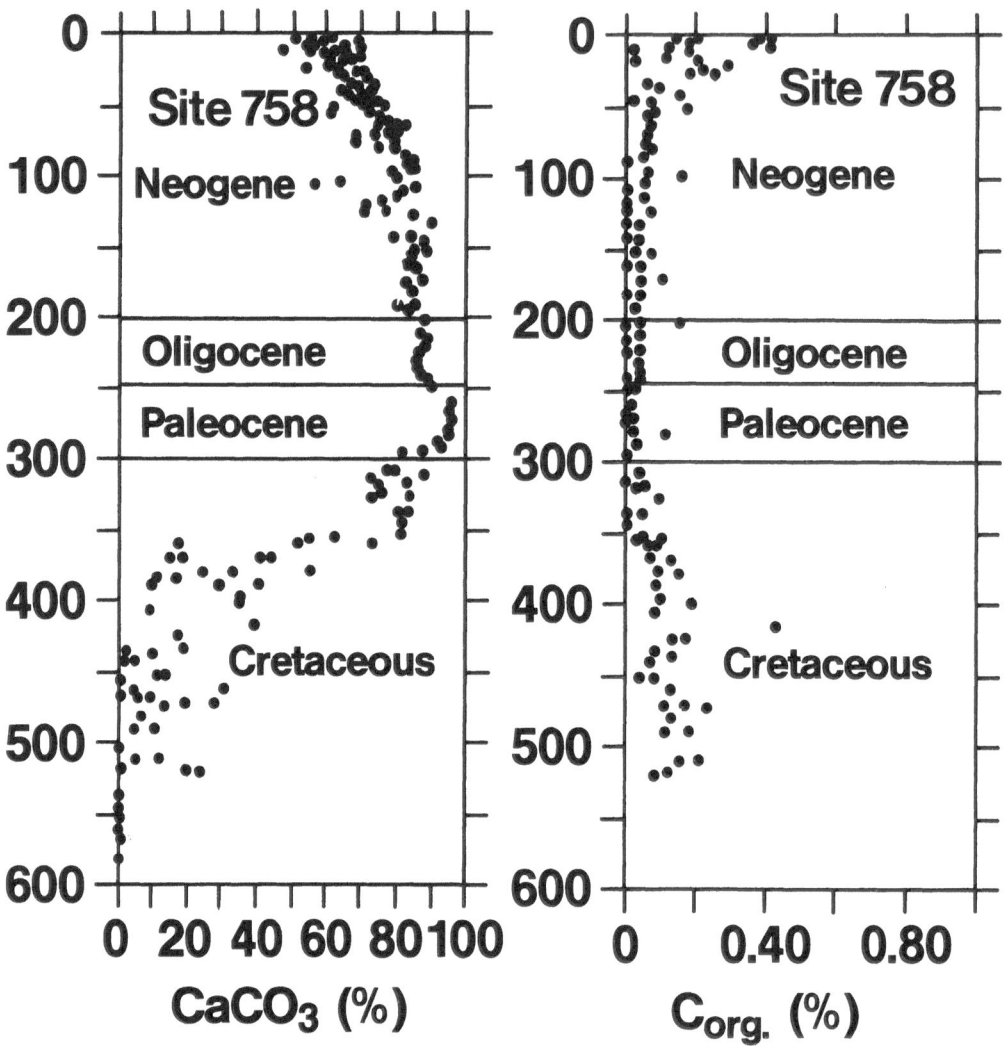

Fig. 14:   Carbonate and organic carbon percentages in Cretaceous to Holocene sediments of ODP Site 758, northern Ninetyeast Ridge, Indian Ocean. The decrease in carbonate and increase in C_{org} in the upper 100m is an effect of the enhanced clastic supply by the Bengal fan (Ganges and Brahmaputra) which itself is a response to the ongoing rise of the Himalayan mountains. In other words, the source for organic particles became more proximal with increasing progradation of the Bengal fan. Higher C_{org}-values in the Cretaceous are affected by supply of terrigenous organic particles in the smaller Cretaceous Indian Ocean (from Littke et al., 1991c). Eocene sediments are eroded.

	Distance (km)	Water depth (m)	C_{org} (%)	HI (mg hc/g C_{org})	Origin
Site 615 (Lower Mississippi Fan)	450	3284	0.7	76(89)	Terrestrial (vitrinite)
Site 620 (Middle Mississippi Fan)	250	2612	1.0	98(107)	Terrestrial (vitrinite)
Site 720 (Middle Indus Fan)	850	4045	0.7	52(108)	Terrestrial
Site 721 (Distant Indus Fan + Upwelling Deposits)	850	1945	1.0	189(400)	Marine (+ Terrestrial)
Site 758 (Distant Bengal Fan)	1800	2924	0.4	45(78)	Terrestrial (inertinite)

Table 7:   Organic matter characteristics of progradational submarine fan deposits. Distance = distance between river (delta) mouth and depositional site; Water depth = present water depth; C_{org} = mean organic carbon percentage; HI = mean and maximum (in parentheses) hydrogen index value; Origin = origin of organic particles based on microscopy and/or geochemical results. Sites 615 and 620 according to Marzi and Rullkötter (1986); Site 720 (upper 150m, Pleistocene) and 721 (upper 100m, Pleistocene and Late Pliocene) according to Prell, Niitsuma et al. (1989), Site 758 (upper 30m , Pleistocene and Late Pliocene) according to Peirce, Weissel, et al. (1989).

Results on organic matter characteristics in the middle Indus fan (ODP Site 720, Fig. 10; Prell, Niitsuma et al., 1989) show strong similarities to the results obtained on the Mississippi and Bengal fans (Table 7). C_{org}-values generally vary between 0.5 and 1% and the organic matter is hydrogen-poor (HI-values less than 110) as typical for terrigenous input. Here, rapid sedimentation by turbidity currents (Prell, Nittsuma et al., 1989) probably enhanced preservation of terrigenous organic matter. At the adjacent ODP Site 721 on Owen Ridge, which is overlain by an upwelling area, turbidites of the Indus fan did not reach the site of deposition and organic matter is mainly of marine origin (Prell, Niitsuma, et al., 1989).

In summary, submarine fans are characterized by higher C_{org} concentrations than most other deep sea sediments. The organic matter is, however, generally hydrogen-poor and mainly of terrigenous origin. Its character seems to be strongly influenced by transport distance and intensity. Preservation of organic matter is enhanced by rapid turbidity sedimentation. It should be noted that locally favourable conditions for the preservation of marine organic matter may occur within submarine fans, e.g., due to the presence of an ideal topography or hypersaline brines (see Kennicutt II, 1986).

3.7   Evaporitic Environments

Relatively few detailed geochemical and petrographic studies until recently dealt with the organic matter in evaporites and carbonates. This is surprising in view of the fact that biomass production can be extremely high in evaporitic environments, e.g., 1810 mg C_{org}/m^2/day in Solar Lake (Cohen et al., 1977) and degradation of organic matter by methanogenic bacteria is inhibited (Hite and Anders, 1991). As carbonates are deposited in many different environments, the following discussion is restricted to three evaporitic sequences in which chemical precipitation of carbonate and other minerals occurs or occurred.

Kenig et al. (1990) and Kenig and Huc (1990) described organic matter from modern carbonates deposited in a hypersaline lagoon in Abu Dhabi. The lagoon provides three principal sources of sedimentary organic matter: microbial mats, mangroves, and seagrass (Fig. 15). The latter grows in the central part of the lagoonal area, where C_{org} values reach 1.3% in surface sediments. Decreasing C_{org} values with increasing depth could indicate that preservation of seagrass in the upper centimetres of the sediments is poor, but other explanations are also possible (Kenig et al., 1990). Mangroves grow at shallower water depth in the intertidal zone (Fig. 15). One mangrove soil contains 8.2% C_{org} and consists of root and leaf (cuticle) tissues (Kenig and Huc, 1990). At greater depth, C_{org} values are lower (0.5-4.6%; Table 8). Organic matter in the supratidal sabkha environment is derived from microbial mats. Corresponding sediments consist of interlayered detrital (storm) deposits and organic mats with C_{org} values between 1.6% (seaward) and 4.5% (landward; Kenig and Huc, 1990). The organic matter of all three principle sedimentary environments is hydrogen-rich (HI=400-750).

Whereas C_{org}-rich sediments accumulate in the sabkha and intertidal zone of the Abu Dhabi lagoon, the adjacent shallow marine carbonates (ooliths) are poor in organic carbon (0.1%, Kenig et al., 1990). Higher C_{org}-percentages (0.2-0.4%) and considerable hydrocarbon quantities were reported for ooliths from the British Jurassic (Ferguson, 1987), and much higher C_{org}-values were found by Ferguson and Ibe (1982; 2.3% C_{org}) for recent ooids. At least with respect to high C_{org}-values in ancient oolithic limestones, it has to be taken into account that (part of) the organic matter can be an oil residue, i.e., oolithes can serve as petroleum reservoirs (see Chapter 6.2).

As another type of a recent evaporitic deposit, Barbé et al. (1990) studied a modern Spanish saline (Fig. 16). The carbonates deposited in the pond at relatively low levels of evaporation contain 3.1-3.4% C_{org}. Sulfates and halite, precipitated at higher levels of evaporation, contain much less organic carbon (0.5 and 0.25%, respectively). This is mainly a dilution effect. Accumulation rates of organic carbon are high for sulfates and halites, because these sediments are deposited at extremely high sedimentation rates and form the bulk of the total evaporitic sequence. Barbé et al. (1990) demonstrated that the molecular composition of compound classes (sterols, lipids, hydrocarbons) differs considerably between the different facies as a response to the variability of produced biomass. For example, only the hydrocarbon fractions of carbonates are characterized by a dominance of highly branched acyclic isoprenoids, which may be derived from diatoms or green algae (see Table 8, Fig. 16). These compounds are absent or very minor in gypsum, which in turn contains abundant sterenes.

Locality	Lithology	C_{org}	Geochemical Data
Hypersaline Lagoon[1] (Abu Dhabi)	Lagoonal mats (seagrass)	0.3-1.0	HI=550-710
	Coastal soils (mangroves)	0.5-4.6	HI=400-550
	Sabkha (microbial mats)	0.5-2.7	HI=510-675
	Shallow marine oolithes	0.1	-
Modern saline[2] (Spain)	Carbonate	3.1-3.4	Highly branched acyclic isoprenoids C_{15}-C_{18} and C_{25}-C_{33} n-alkanes
	Gypsum	0.5	
	Halite	0.25	Dihydrophytol, phytanic acid
Oligocene evaporites[3] (France)	Marlstone	0.5 - 5.0	HI=230-580
	Anhydrite	< 0.5	-
	Halite	< 0.5	-

Table 8:   Organic matter characteristics of three evaporitic deposits.
[1]: after Kenig et al, 1990;
[2]: after Barbé et al., 1990;
[3]: after Blanc-Valleron et al.,1991

Ancient evaporitic sediments of Oligocene age from the Mulhouse basin (France) were studied in detail by a joint European organic geochemical research group. First results (Blanc-Valleron et al., 1991) reveal that halites and anhydrites are poor in organic matter (less than 0.5%; Table 8), but that organic matter accumulation rates were high due to high sedimentation rates. Interlayered thin marlstones do, however, contain up to 5% C_{org}, and the organic matter is hydrogen-rich (H/C=0.9-1.5, HI=230-580). Saturated hydrocarbons are characterized by high concentrations of phytane, pristane and other isoalkanes. Interestingly, the yield of C_{org}-normalized solvent extract from these evaporitic rocks is extremely high, if compared to other immature sediments (Fig. 17; Hofmann, 1992).

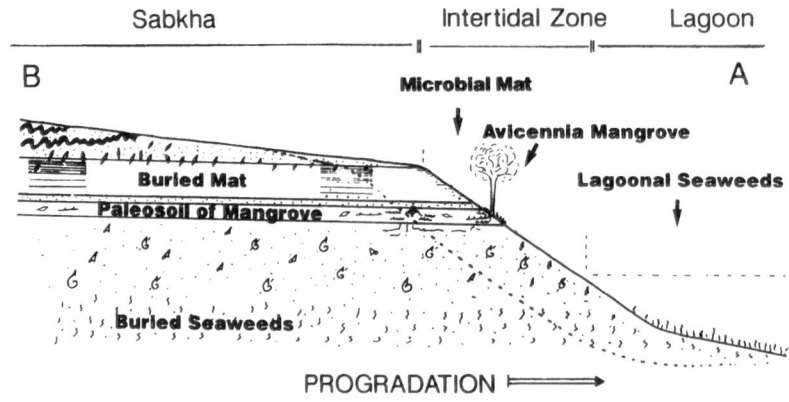

Fig. 15:   Schematic section through the Abu Dhabi lagoon coast line (from Kenig and Huc, 1990; see text for geochemical characteristics).

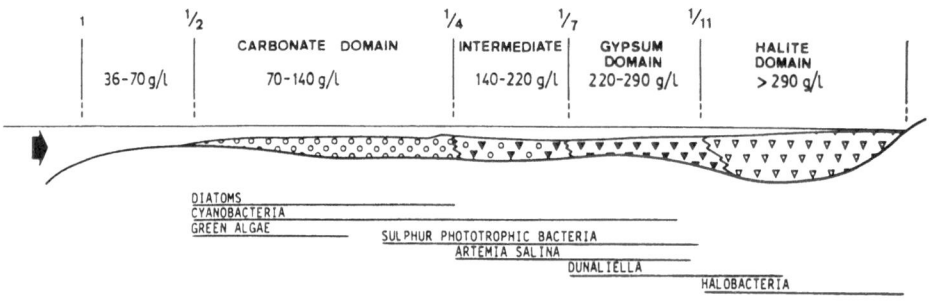

Fig. 16:   Schematic section through a modern saline in Spain with evaporation ratios, salt concentrations and occurrence of different groups of biota (after Barbé et al., 1990).

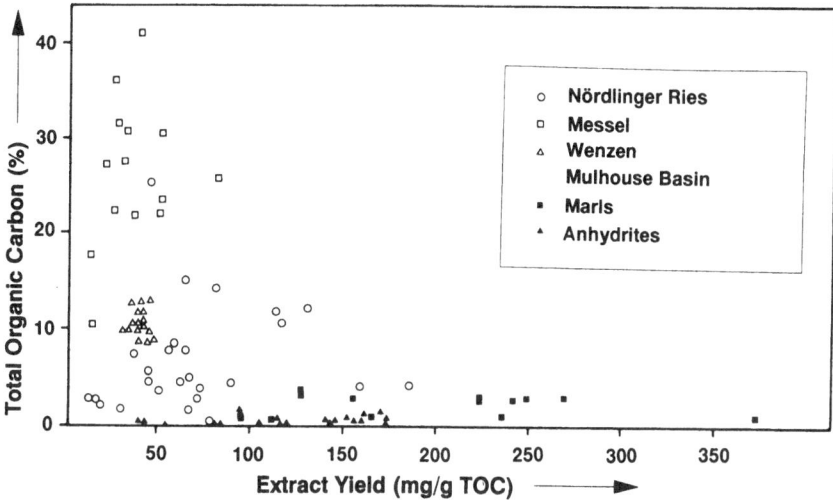

Fig. 17:   Organic carbon percentages (C_{org}) versus solvent extract yields of samples from an evaporitic environment (Mulhouse basin, S-bed) in comparison to other immature oil shales (from Hofmann, 1992 after Littke et al., 1988).

In summary, in evaporitic sequences high organic matter concentrations can be expected only in carbonates and anhydrites, as well as in intercalated shales. In halites and potassium salts, C_{org}-values are low due to the dilution of organic matter by rapid precipitation of minerals. Organic matter in evaporitic sequences is often hydrogen-rich; this may be due to an inhibited degradation before and after deposition. Geochemically, differences between various evaporitic settings are more obvious than similarities. Even relatively well-established organic indicators of highly saline depositional environments such as low pristane/phytane ratios (ten Haven et al., 1988), predominance of even numbered over odd numbered n-alkanes (Welte and Waples, 1973), and the occurrence of gammacerane (Mello et al., 1988) do not seem to be of general validity (de Leeuw and Sinninghe Damsté, 1990).

3.8   Lakes

Lakes are known to be suitable settings for deposition of rich petroleum source rocks. According to the kerogen classification scheme by Espitalié et al. (1977), the most hydrogen-rich type I kerogens (H/C ca.

1.5 and more, O/C smaller than 0.1; Fig. 2) are mainly derived from lacustrine depositional environments (Tissot and Welte, 1984).

The organic richness of lake deposits is extremely variable depending, e.g., on the nutrient supply by inflowing rivers, water circulation in the lake and the stability of water stratification. A mapping of organic richness based on organic carbon measurements on recent sediments from Lake Tanganyika was presented by Huc (1988a,b). His results demonstrated that - similar to the situation in silled marine basins such as the Black Sea - the most organic matter-rich sediments were deposited in the deepest part of the lake. The distribution pattern is, however, complex, testifying that not only water depth influences organic richness.

The Eocene Messel Shale serves as an interesting example. In these sediments organic carbon values vary between 13% and 39% (Rullkötter et al., 1988b; Table 9). Careful palaeoenvironment reconstructions indicate that this spectacular organic richness in the 200 m thick sequence is the consequence of the combination of ideal environmental factors. A warm and humid climate and an inflowing river with nutrients fostered plant growth at the lake margins and in the photic zone of the lake (Franzen and Michaelis, 1988). On the other hand, river in- and outflow did not cause water circulation to an extent that water stratification was inhibited. This is obvious from the rather regular interlayering of algal-rich laminae (often about 0.01 mm thick) and clay-rich laminae (often about 0.1 mm thick; Plate 1, C and D) which indicates a seasonal deposition typical of stratified lakes. The lack of oxygen at the sediment/water interface is also indicated by absence of bioturbation. Furthermore, the tectonic setting of the lake in the developing graben system of the Upper Rhine valley permitted deposition of a thick lacustrine sequence.

Locality, Age	Lithology	C_{org}	S	HI (mg hc/g C_{org})
Messel, SW-Germany (Eocene)	finely laminated, fine-grained shales	28.3±6.1[1] min. 7 max. 40	1.0±0.6 min. 0.2 max. 2.4	553±43[1] min. 470 max. 640
Nördlinger Ries, Southern Germany (Miocene)	extremely variable[2] finely laminated shales, marlstones, dolomites and others	6.9±5.3[3] min. 1.6 max. 25	2.4±1.2 min. 0.5 max. 4.8	513±176[3] min. 177 max 920

Table 9:   Organic matter characteristics of deposits from an Eocene freshwater lake (Messel) and a Miocene hypersalinar lake (Nördlinger Ries).
[1]   Rullkötter et al. (1988b);
[2]   Jankowski (1981);
[3]   Rullkötter et al. (1990).

A different pattern of C_{org} contents characterizes the Nördlinger Ries Shale (Rullkötter et al., 1990) which was deposited as a lacustrine filling of a meteoric impact crater of Miocene age. In contrast to the Messel Shale its filling history was not influenced by tectonic processes, i.e., the basin became shallower with increasing thickness of the lacustrine deposits. C_{org}-values vary between less than 2 and 26% in this sequence (Table 9), which reaches a maximum thickness of more than 200 m in the central part of the crater. Organic matter-rich intervals (C_{org} greater than 8%) are relatively rare and restricted to several dark green layers, which are usually less than 10 cm thick. These layers are concentrated in the lower half of this sequence, testifying that a critical water depth of about 100m was necessary to allow deposition of organic carbon-rich sediments. Also, great differences between C_{org}-values in marginal and central parts of the basin were found (Rullkötter et al., 1990).

The type of organic matter deposited in the two lakes is also different and clearly depends on the environmental condition in the hinterland. In the case of the Messel Shale, remains of higher land plants were fluvially transported into the lake. Accordingly, the Messel Shale kerogen is composed of about 20% terrigenous higher plant material and 80% autochthonous material derived from subaquatic organisms (Jankowski and Littke, 1986; Rullkötter et al, 1988b). In the Nördlinger Ries Shale, terrigenous organic particles such as vitrinite and inertinite are generally less abundant, especially in the central part of the deposit (Rullkötter et al., 1990). This is probably due to the lack of an effective transport mechanism, i.e., no river was flowing through the lake as in the case of the Messel lake. Therefore, terrigenous particles were mainly deposited at the rims of the basin.

Another important environmental factor influencing kerogen composition is water chemistry. The Ries lake had variable, often anomalously high salinities (Jankowski, 1981). Obviously, organic matter input into the sediments changed with these changes in salinity, leading to the great variability in organic petrographic and bulk geochemical characteristics, e.g., in Hydrogen Index values (Table 9). The fact that the three most hydrogen-rich samples of the Nördlinger Ries Shale (HI greater than 800) are characterized by a fluorescing, structureless organic groundmass (Rullkötter et al., 1990) indicates an origin from bluegreen algae, which are relatively resistant to hypersaline conditions (Jankowski, 1981: 135; Barbé et al., 1990). No salinity changes influenced the Messel Shale, which contains a remarkably homogeneous kerogen on a petrologic and bulk geochemical level. Its rather uniform, intermediate HI-values (Table 9) formally indicating type-II kerogen are the effect of a mixture of aqueous (type-I) and terrigenous (type-III) organic matter.

Also, ionic composition of the water in the two lakes was very different: Lake Messel was rich in iron, whereas the Ries lake, completely surrounded by Malmian carbonates, was dominated by alkaline and alkaline earth ions (Na, K, Ca, Mg) and poor in iron (Jankowski, 1981). This difference in iron availability greatly influenced the sulfur content of the organic matter. In the Messel lake, little sulphate was available and sulfur was mainly fixed in pyrite. In the Ries lake, more sulphate was available (Fig. 18) and fixation of sulfur in pyrite was inhibited due to the low iron concentration. A great part of the available sulfur was therefore bound in organic matter, leading to anomalously high atomic S_{org}/C_{org} ratios (up to 0.08, Rullkötter et al., 1990). The Nördlinger Ries data show that hypersaline lakes may produce sediments as rich in sulfur as marine sediments. In this case, the sulphur versus C_{org} plot (Fig. 18) cannot be used to differentiate between marine and lacustrine sediments (Berner, 1984).

In the Nördlinger Ries deposits, extract yields normalized to organic carbon are much higher than would be expected for immature organic matter at very shallow depth (Fig. 17). In most samples, phytane is the most abundant single aliphatic compound and phytane/pristane ratios are extremely high (greater than 10) as typical of sediments deposited under hypersaline conditions (ten Haven et al., 1988). The "aromatic hydrocarbon" fraction is dominated by thiophenes, thiolanes, and other unidentified sulfur compounds (Rullkötter et al., 1990), thus proving the early diagenetic sulfur incorporation into the kerogen. The high sulfur content of the organic matter in the Nördlinger Ries sediments is believed to cause early petroleum generation, i.e., petroleum generation at very low temperatures, because sulfur-carbon bonds are more easily cracked than carbon-carbon bonds (Baskin and Peters, 1992). Similar conclusions were drawn for the diatomaceous Monterey formation in California which is also characterized by sulfur-rich organic matter (Baskin and Peters, 1992). In the case of the Nördlinger Ries kerogen, high sulfur contents do not only cause high extract yields, but probably also pyrolysis at low temperatures leading to two more or less separate S_2-peaks and to very low T_{max}-values (see Fig. 4) in several samples.

In summary, lakes are potential settings for organic matter-rich sediments with hydrogen-rich kerogen. Due to their smaller size compared to most marine basins, their deposits are strongly influenced by environmental factors such as river inflow and outflow and water chemistry. Accordingly, variability of organic richness, kerogen type, and bitumen yield and composition are great both in vertical sections and horizontal profiles.

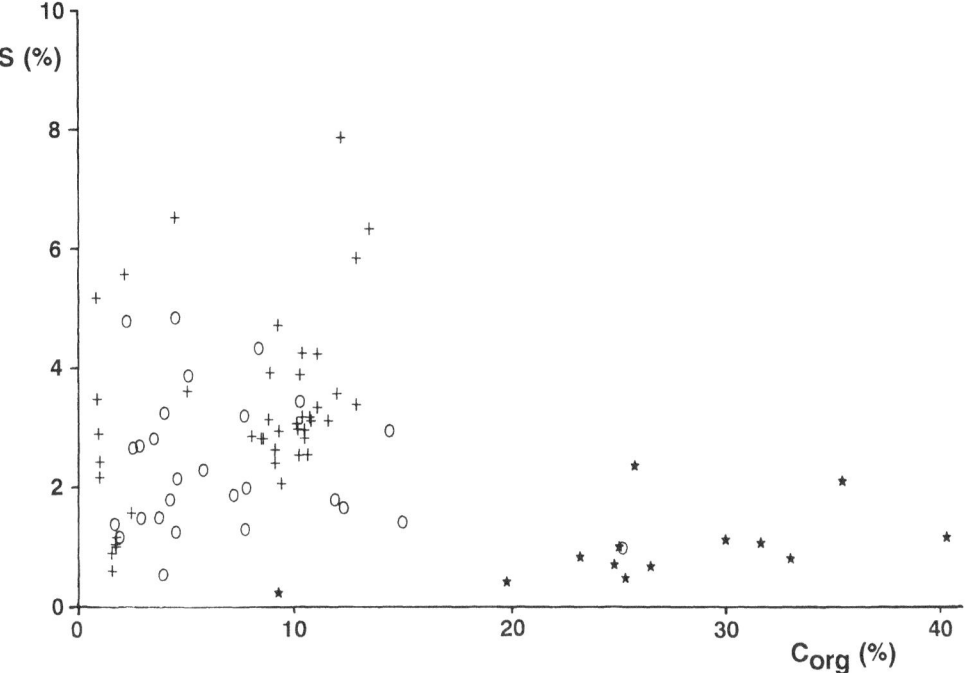

Fig. 18: Organic carbon percentages (C_{org}) versus sulfur percentages for three different, immature oil shales. Crosses mark the Lower Toarcian Posidonia Shale from the Wenzen borehole, circles the hypersaline, lacustrine Nördlinger Ries deposit of Miocene age, and stars the Eocene, lacustrine Messel Shale.

3.9   Fluviodeltaic coal-bearing strata

Fluvial and deltaic sequences are highly variable with respect to concentration of organic matter. This heterogeneity is partly due to the fact that preservation of organic matter is strongly affected by differences in climate and exposure to either subaeric or subaquatic conditions after deposition. In fluvial and deltaic sequences, the bulk of the organic matter consists of parts of higher land plants most of which are deposited close to but not exactly at the place of plant growth (Scheihing and Pfefferkorn, 1984). As growth of higher plants is inhibited in arid regions, deposition of organic matter-rich fluvial and deltaic sediments is usually restricted to humid climates. Furthermore, source rocks are not expected to occur in pre-Devonian fluviodeltaic rocks, because the evolution of higher land plants only started during the late Silurian.

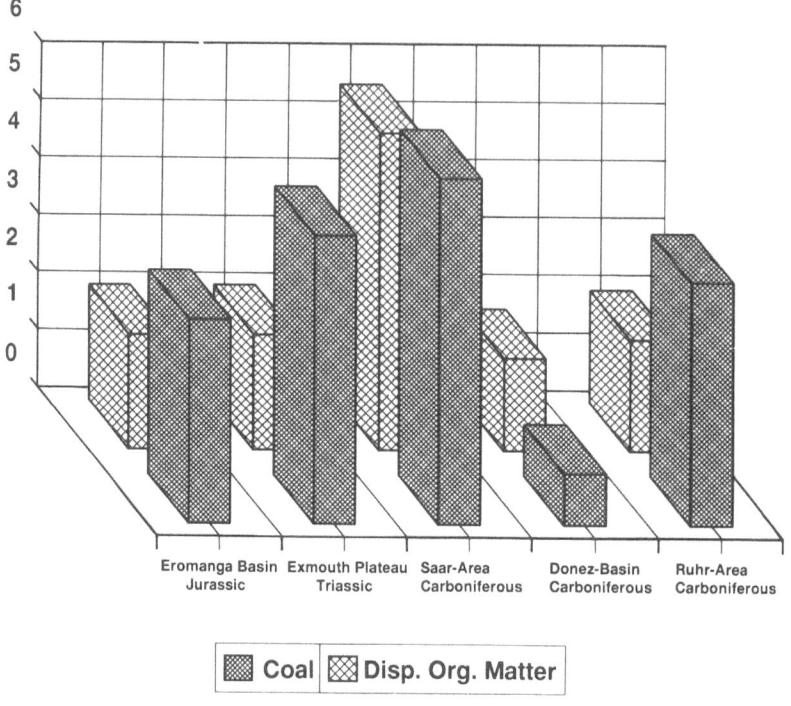

Fig. 19: Volume percentages of organic matter bound as coals seams and as dispersed organic matter in different coal-bearing basins (compilation by Scheidt and Littke, 1989 based on literature data).

Coal-bearing fluvial and deltaic sequences deposited in humid climates contain the highest concentration of organic matter in rocks on a meter scale (50-99% C_{org} in coals), and on a basin-wide (kilometer) scale. In Fig. 19 organic matter percentages are compared for five coal-bearing basins according to a compilation by Scheidt and Littke (1989). In most of these basins, more organic matter is fixed in coals than finely dispersed in clastic sediments. The average amount of total organic matter ranges from 4 to 12 vol.% and the ratio of coal to dispersed organic matter from 0.7 to 2.5% over intervals of several hundred to more than a thousand metres thickness.

Total amounts of dispersed organic matter are different in different lithologies as exemplified for the Ruhr basin in Fig. 20 (Scheidt and Littke, 1989). For this area, a decrease in organic matter content with increasing grain size was found, e.g., mudstones on an average contain more organic matter than siltstones and sandstones. More than 2 vol.% dispersed organic matter on an average is fixed in gray mudstones, one of the major lithologies. In contrast, the various siltstone and sandstone facies contain less than 2 vol.-% and often only 1 vol.-% dispersed organic matter. The only exception are rare conglomeratic sandstones with more than 2 vol.-% dispersed organic matter in the form of coal pebbles and barks of twigs and stems. Huc et al. (1986) reported similar values for shale samples from other coal-bearing sequences, i.e., the Cretaceous Douala basin (1-3% C_{org}), the Miocene Mahakam delta (1-4% C_{org}) and the Westphalian in the Paris basin (1-5% C_{org}). Few clastic rocks in coal-bearing basins contain more than 10% C_{org}. Exceptions seem to be black shales (see Fig. 20) derived from marine ingressions (Wenger and Baker, 1986: up to 20% C_{org}) and mudstone partings within coals seams (Littke and ten Haven; 1989: up to 32% C_{org}). The organic richness of the former group is explained by high algal production and the occurrence of humic precipitates fostered by the flooding of swamps (see Chapter 3.5), whereas the richness of the latter is partly due to the in-situ growth and preservation of roots after clastic deposition (Scheidt, 1988).

In fluvial and deltaic sequences, terrigenous organic matter usually predominates over aquatic organic matter (e.g., Scheihing and Pfefferkorn, 1984; Scheidt and Littke, 1989; Smyth, 1989), although the latter occurs in interbedded 1) transgressive marine deposits (e.g., Wenger and Baker, 1986) and 2) lacustrine deposits such as rare sapropelic coals (e.g., Stach et al., 1982). Terrigenous organic matter is known to contain less hydrogen and more oxygen than aquatic organic matter (Pelet, 1983; Tissot and Welte, 1984). Accordingly, in immature fluvial and deltaic rocks, H/C ratios of kerogen are generally lower than 1 and O/C ratios greater than 0.2 (Boudou et al., 1984; Huc et al., 1986; see Fig. 2). Both ratios decrease with increasing maturity (van Krevelen, 1961; Boudou et al., 1984; Littke et al., 1989).

	Area	Age	R_r (%)	C_{org} (%)	HI (mg hc/g C_{org})	Major Macerals
Talang Akar coals	Java	Tertiary	0.3-0.8	59.9	299(374)	V>L>I
Malagasy coals	Madagascar	Carboniferous-Permian	0.7-0.8	59.9	182(284)	V~I>L
Malagasy mudstones	Madagascar	Carboniferous-Permian	0.7-0.8	14.4	143(250)	I~L>V
Ruhr coals	Germany	Carboniferous	0.7	75.2	234(315)	V>I>L
Ruhr mudstones and siltstones	Germany	Carboniferous	0.7	6.6	93(191)	V>I>L

Table 10:   Organic matter characteristics of coal-bearing strata (coals, mudstones, siltstones) after Horsfield et al. (1988), Ramanampisoa et al. (1990), and Littke et al. (1989). R_r = mean vitrinite reflectance, HI = Hydrogen Index, mean and maximum value (the latter in parentheses). V = vitrinite, I = inertinite, L = liptinite.

Hydrogen Index values of coals are generally in the range of 150 to 300 at immature or marginally mature stages (Table 10); similar values are typical for handpicked vitrinites from coals (Littke et al., 1988) and suggest an intermediate hydrocarbon generation potential. Interestingly, HI-values of kerogen in mudstones and siltstones interbedded with the coals are generally lower (Table 10), although the petrographic composition of the organic matter is roughly similar (Scheidt and Littke, 1989; see Chapter 7 for explanation).

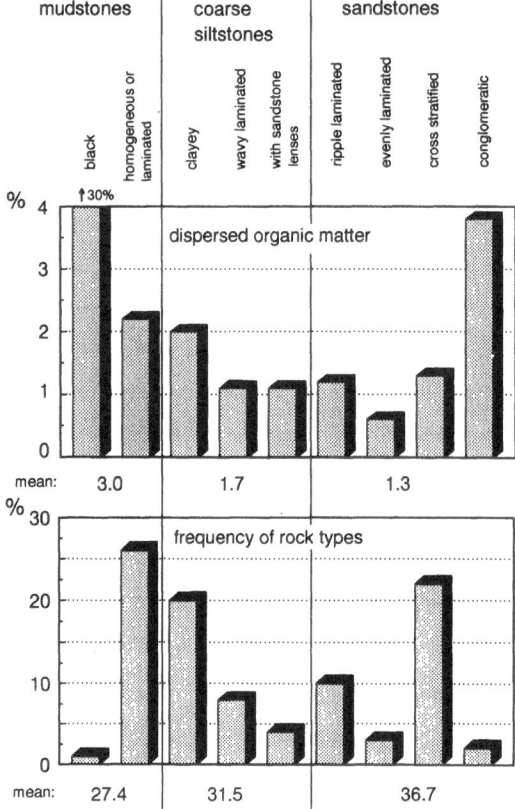

Fig. 20:   Volume percentages of organic matter in different clastic rocks of the Carboniferous Ruhr basin (upper part) and percentage of these rock types in this sedimentary sequence (lower part). The grain size of the sediments as well as the average grain size of the organic clasts is increasing from left to right (after Scheidt and Littke, 1989). The volume percentage of organic matter in black shales is off scale (about 30%).

Petrographically, organic matter in fluvial and deltaic deposits is usually composed of a mixture of different macerals. Vitrinite is the most frequent maceral group in most coals (Stach, 1982), although in several coal-bearing strata such as the Permian basins of the southern hemisphere inertinite prevails (Chandra and Taylor, 1982; Smyth, 1989; Ramanampisoa et al., 1990). Liptinite rarely forms more than 50% of coals, although there are exceptions (e.g., Horsfield et al., 1988). Figure 21 provides a compilation of average vitrinite, inertinite, and liptinite percentages of coals from different basins. Average liptinite contents vary between 5 and 35%, average vitrinite contents between 34 and 88%, and average inertinite contents between 1 and 50%. As "the chemical and botanical precursors of inertinite are mainly the same as those for vitrinites, namely cellulose and lignin from the cell walls of plants" (Teichmüller, 1982) and vitrinite is thought to reflect better preserved organic matter than inertinite, the ratio vitrinite/inertinite is used as an indicator for the degree of pre- and syndepositional degradation of plant particles (Shibaoka and Smyth, 1975; Diessel, 1987; Littke, 1985, 1987).

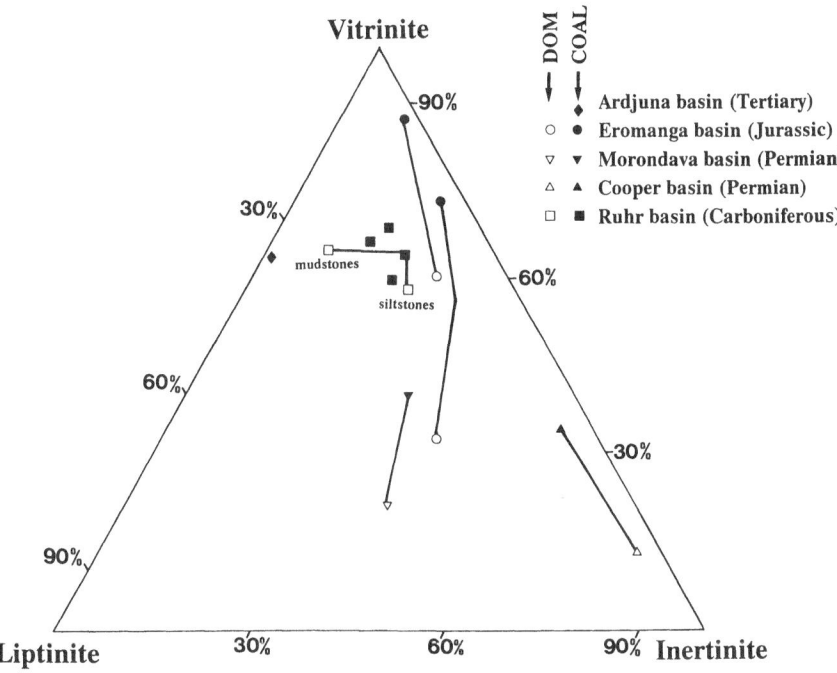

Fig. 21.   Average vitrinite, inertinite, and liptinite percentages of coals and corresponding kerogen in clastic rocks (dispersed organic matter; DOM) in different coal-bearing basins.

The petrographic composition of organic particles in clastic rocks is either similar to that of associated coal seams (Scheidt and Littke, 1989) or vitrinite is less abundant (Smyth, 1989; Ramanampisoa et al., 1990). In most basins, inertinite seems to be enriched in clastic rocks (Fig. 21) relative to coals. In the Ruhr basin, this enrichment is less obvious than in other basins and was only established for siltstones, whereas mudstones are enriched in liptinite on an average. This is explained by transport sorting leading to a relative enrichment of the lightest maceral group liptinite in mudstones deposited at lowest current velocities. The predominance of vitrinite is explained by the great percentage of vitrinite roots in the clastic rocks of the Ruhr basin. It should, however, be noted that the variability of petrographic composition in each rock type is enormous, e.g., liptinite percentages (vol.% of organic matter) range from 0 to 60% in mudstones and siltstones and from 0 to 20% in sandstones (Scheidt, 1988).

4.   DEPOSITIONAL HISTORY OF THE POSIDONIA SHALE

4.1   Overview

The Lower Toarcian Posidonia Shale (PS) is one of the most widespread and economically important petroleum source rocks of Central and Western Europe, e.g., sourcing most of the oil accumulations in northern Germany (Schwarzkopf and Leythaeuser, 1988; Wehner et al., 1989), in the Paris basin (Espitalié et al., 1987) and in parts of the Upper Rhine Graben (Welte, 1979). The following chapter will present a summary of sedimentological, petrological and geochemical data on immature PS cores from the Hils syncline in northern Germany (Littke et al., 1991b) and from the Schwäbische Alb in southern Germany (Littke et al., 1991a) and discuss these data to elucidate the depositional history. In northern Germany, the PS covers the same three ammonite zones as in southern Germany (Riegraf et al., 1984; Loh et al., 1986), i.e., *tenuicostatum*, *falciferum*, and *bifrons*. As the duration of these ammonite zones is estimated to be about 1 million years (Hallam, 1975), a total duration of the PS deposition of two to three million years can be expected. The duration of the entire Lower Toarcian is three to four million years according to geological time tables (Harland et al., 1990). This information allows the calculation of accumulation rates for organic matter and for minerals of the PS.

The samples from the Hils syncline were taken from six shallow boreholes (Fig. 22) from which complete cores of the PS were obtained. In each of these cores the PS is overlain by more than ten metres of Dogger sediments. The discussion of organic geochemical data will be restricted to the most immature cores Wenzen and Wickensen, which did not experience severe changes of the organic matter by thermal maturation (Leythaeuser et al., 1988; Littke et al., 1988; Rullkötter et al., 1988a; Schaefer and Littke, 1988). One hundred and two samples from the Wickensen core served as the principal subject for the geochemical and petrographic study, because this core includes a thicker and more complete PS section than the Wenzen core.

In southern Germany, a total of 224 core samples were taken from six shallow boreholes along the Schwäbische Alb (Fig. 23, Rotzal, 1990; Bauer, 1991). In southern Germany as well as in the Hils syncline, initial samples were taken roughly every metre or every 50cm along the cores. This was followed by detailed sampling, often closely spaced, which was concentrated across important lithologic boundaries and where the initial analytical results indicated possible anomalies.

47

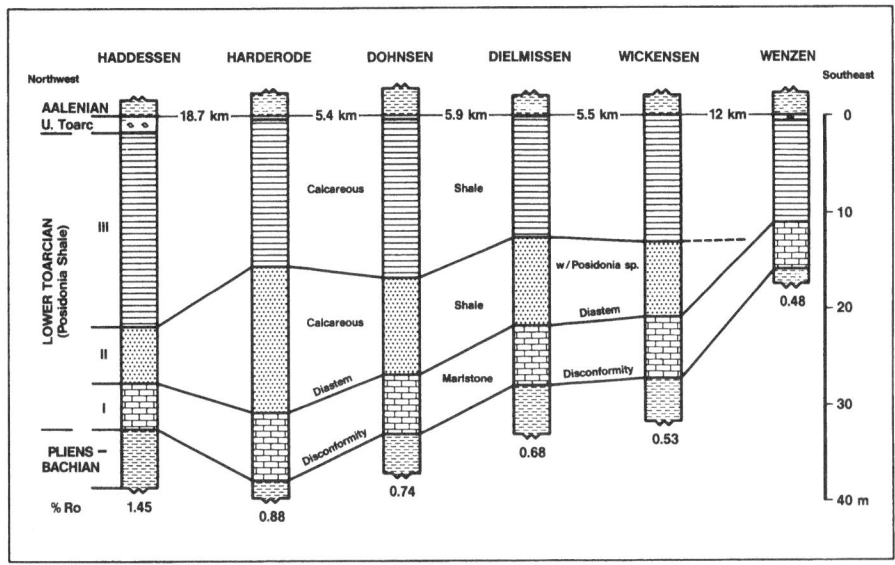

Fig. 22: Stratigraphic profile of PS cores in the Hils area from Littke et al. (1991b). Thickness is corrected for dip but not decompacted. A detailed map of the Hils area with location of boreholes is published in Littke and Rullkötter (1987). The marlstone (ms) corresponds to Unit I in the text, the Posidonia zone (pz) with abundant bivalves to Unit II, and the calcareous clay shale (c.cl.sh) to Unit III. Vitrinite reflectance (R_o) increases from SE (0.48 %) to NW (1.45 %).

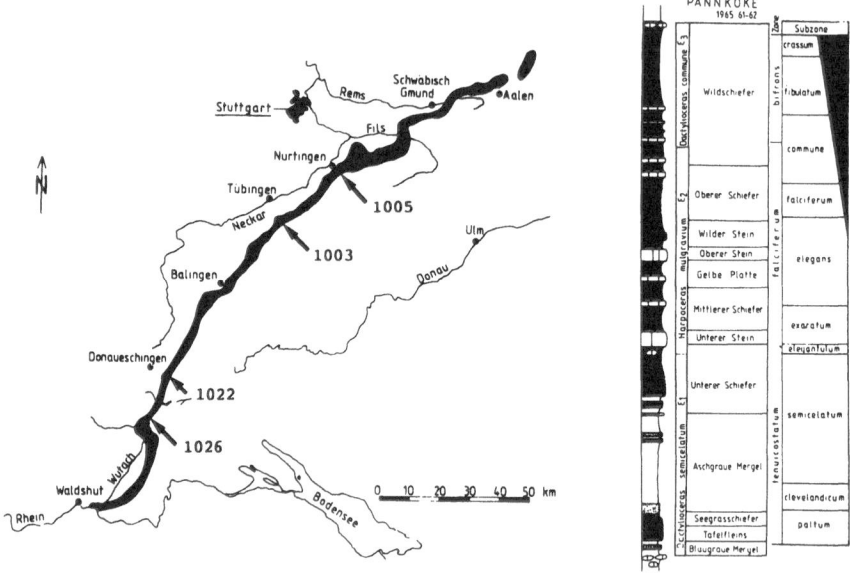

Fig. 23:   Location of boreholes 1003, 1005, 1022, and 1026 in the Schwäbische Alb, southern Germany (from Littke et al., 1991a). Boreholes 1023 and 1025 which were also studied but which are not discussed here in detail are close to 1022 and 1026, respectively. The right part of the figure shows a schematic stratigraphic column of the Lower Toarcian in southern Germany.

4.2   Lower Toarcian black shales in Europe

The distribution of Toarcian black shales in central and northern Europe is shown in Fig. 24. At the same time, black shales were also deposited in the Tethyan ocean (Baudin et al., 1988, 1990). Both areas were interconnected as evident from the occurrence of a tropical Tethyan fauna in the German PS (Riegraf, 1985).

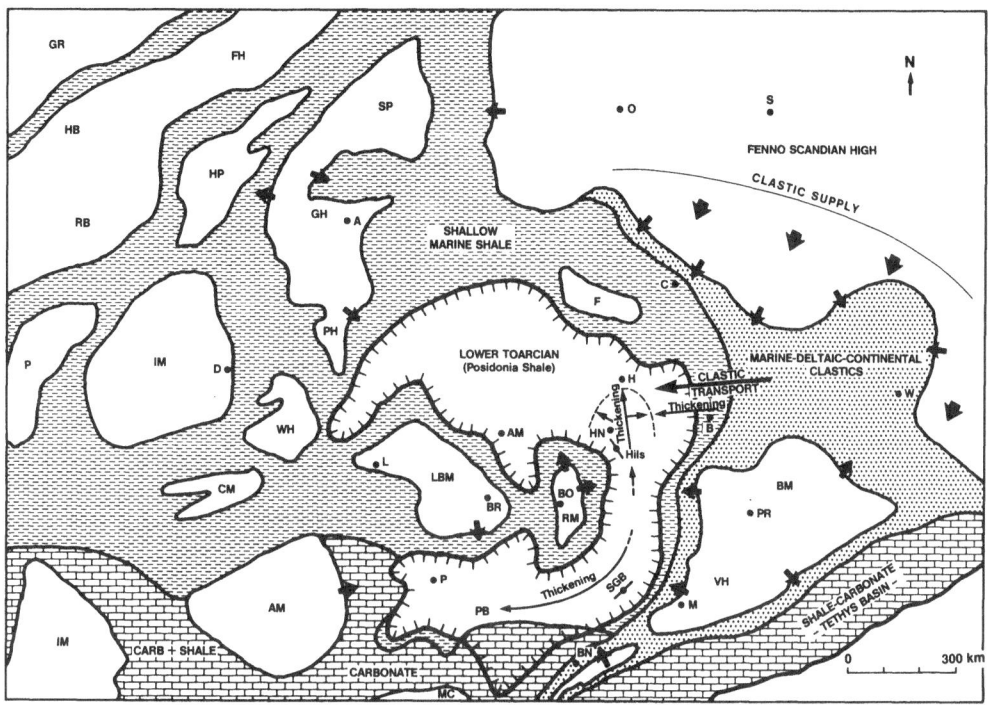

Fig. 24:   Sinemurian - Aalenian paleogeographic map from Littke et al. (1991b) generalized
and sketched from Ziegler (enclosure 18, 1982) showing major emergent
(nondeposition) areas and general distribution patterns of principal sedimentary
facies. The outline of the PS is only for the principal 'central' region of its deposition
in western Europe; similar facies of this age are known beyond the outlined area; the
PS boundary was drawn to conform with distribution patterns shown by Riegraf
(1985) and Loh et al. (1986). The map indicates that major supply and transport of
clastics was from the northwest and east from the Fenno Scandian High and
contiguous areas. Thickening and thinning lines refer to isopachous variations of the
PS. The European area was open to the Tethys ocean at various places during the
Lower Jurassic (including Lower Toarcian). Location of the Hils area is shown.
Codes: Emergent areas: AM, Amorican Massif; BM, Bohemian Massif; F, Fünen
High; GH, Grampian Highlands; LBM, London-Brabant Massif; MC, Massif Central;
PB, Paris Basin; PH, Pennine High; RM, Rhenish Massif; SP, Shetland Platform;
VH, Vindelician High; WH, Welsh High, Cities: A, Aberdeen; AM, Amsterdam; B,
Berlin; BN, Bern; BO, Bonn; BR, Brussels; C, Copenhagen; H, Hamburg; HN,
Hannover; L, London; M, Munich; O, Oslo; P, Paris; PR, Prague; S, Stockholm; W,
Warsaw.

A number of hypotheses about the palaeogeography and palaeoceanography leading to the deposition of
Toarcian black shales have been proposed. For the British Lower Toarcian (Jet Rock), Morris (1979, 1980)
concluded that deposition took place under poorly oxygenated bottom waters and that organic matter was

only affected by slow anaerobic instead of faster aerobic decay (see Daumas et al., 1978; Demaison and Moore, 1980; Deuser, 1974). The detailed sedimentological description by Morris (1980) reveals that these organic matter-rich sediments in Britain are probably less laminated and less calcareous than contemporaneous deposits in northwest Germany (Table 11). A duration of about 0.5 Ma was estimated by Jenkyns (1985) for the entire black shale deposition.

Region	Thickness (m)	Lithology	C_{org} (wt%)	Carbonate (wt %)	HI (mg hc/g C_{org})	Remarks	Ref.
East England and North Sea	10-33	shale	5-15 av.<11.4	2-5	300-600	limestone and shell beds	1
Northwest Germany	16-40	laminated shale and marlstone	av. 11	35-61	600-700	upper shale, lower marlstone units	2
Southwest Germany	5-20	shale and marlstone; partly laminated	9-14 av. 9	av. 40	700	δC^{13}-excursion widespread limestones "anoxic biomarkers"	3 4 5
Paris Basin	10-30 - 65	- - shale-marlstone	5-14 5-11 4-7	10-30 20-40 -	- - 500-700	- homogeneous org. matter -	6 7 8
Tethys	8-80 -	marlstone laminated shale	1-2 av. 1.3	<50 -	200-500 200-300	restricted basins in Greece Italian Alps	9 10

Table 11:   Regional comparisons of the Toarcian black shales of central and northern Europe. References: 1, Jenkyns (1985); Morris (1980); 2, Littke and Rullkötter (1987); Littke et al. (1988); 3, Küspert (1982); 4, Urlichs (1977); 5, Moldowan et al. (1986); 6, Durand et al. (1972); 7, Huc (1977); 8, Espitalié et al. (1987); 9, Baudin et al. (1990); 10, Jenkyns (1985).

According to Hallam (1981) and Jenkyns (1985) the Early Toarcian rise in sea level may have produced an elevated nutrient flux from flooded land areas to the shelf seas which in turn led to a higher productivity at this time. Earlier, Hallam (1967) argued that black shale deposition took place at shallow water depth of only 15-30m, whereas Müller and Blaschke (1969) concluded that the frequent occurrence of coccoliths as major carbonate particles and the similarity to deep water sediments from the Black Sea are indicative of deposition at water depths exceeding 200m.

French geochemists who studied Early Toarican samples from the Paris basin (Durand et al., 1972; Huc, 1977; Espitalié et al., 1987) concluded that the organic matter is remarkably homogeneous, even over great distances (300 km), although there are differences in the composition of the surrounding mineral matter, e.g., in the grain size distribution and carbonate content. The high petroleum formation potential of the Early Toarcian shales was interpreted to be due to favourable conditions for preservation of labile organic matter in an anoxic depositional environment.

In southwest Germany, several models for PS ('Posidonienschiefer', Quenstedt, 1843) deposition were developed. The widely used stagnant basin model (e.g., Seilacher, 1982) was not accepted by Kauffman (1981) who found abundant epibenthic fauna at some levels. He proposed that anoxic conditions were present only within the sediments or in a very thin layer above, but that the bulk of the water mass was oxygenated above an 'algal-fungal' mat that periodically floated a few centimetres above the substrate during black shale deposition (Kauffman, 1979). However, Moldowan et al. (1986) used geochemical arguments such as the low Ni/(Ni+VO) ratios in porphyrins to support an anoxic depositional environment. Küspert (1982) argued that the large $\delta^{13}C$ variation (of organic matter and carbonate) towards isotopically lighter carbon, especially in the lower part of the black shales (his unit A), is the effect of a stagnation-induced carbon recycling. He concluded (Küspert, 1982: 492) that this facies 'coincides with the widest regional occurrence of bituminous rocks in Europe and, hence, marks the period of strongest anaerobism'.

Based on the geographic distribution of Toarcian rocks in southwest Germany, Riegraf (1985) explained the black shale deposition by a transgression during Toarcian time and, as a consequence, less circulation and oxygen supply at the sediment surface due to the deepening of the water in the overall shallow basin. Based on trace element concentration and sulphur isotope data, Brumsack (1991) supported the idea that Toarcian black shales were deposited under permanently anoxic water.

In northwest Germany the Toarcian shales are thicker than in southwest Germany (Schmitz, 1980). Teichmüller and Ottenjann (1977) used samples of PS for the first detailed, modern organic petrological description of organic matter in oil shales. They found that the bulk of the organic matter can be attributed to alginite derived from phytoplankton (green algae) and to bituminite, whereas only small amounts of land plant-derived vitrinites and inertinites are present.

In Toarcian sediments deposited in the Tethyan ocean significantly lower organic carbon values and hydrogen indices (HI) are reported reflecting a lower degree of preservation of organic matter (Jenkyns, 1985, 1988; Baudin et al., 1988; Baudin et al., 1990). Riegel et al. (1986) and Fleet et al. (1987), in extension of the ideas of Jenkyns (1985), interpret the occurrence of oilprone organic matter during Toarcian time as due to an impingement of a Tethyan oxygen-minimum zone on the adjacent epeiric area following a general transgression. According to Parrish and Curtis (1982) seasonal wind-driven upwelling may have affected the northern Tethyan continental margin during the Toarcian. Whether this strong upwelling caused the black shale deposition in central Europe is highly uncertain, because the high degree of water circulation required for this model is probably less favourable for good preservation of organic matter than less intense circulation and watermass stratification (Tyson, 1987).

4.3   Physical stratigraphy and lithology

In the Hils area, the PS (Fig. 22) consists of a lower marlstone (Unit I), a middle calcareous clay-shale characterized by common bivalves (Unit II), and an upper calcareous clay-shale (Unit III). The basal marlstone is a uniform 'blanket' deposit varying between 4.9 and 7.0m in thickness. The combined thickness of the calcareous shale units increases from 10.6 to 30.0m, more than doubling from southeast to northwest (see Fig. 22). This suggests that the distal basinal area was toward the north (see Fig. 24). Unit II is absent at Wenzen. This may be due to depositional onlap (pinchout) over a small topographic high.

Lithological differences among the three PS members are relatively minor. At Wickensen all units are medium-olive gray, rich in calcite and organic matter, fine grained, and well laminated. The main distinction between the marlstone and shales is the carbonate content (Fig. 25A). Carbonate occurs as very thin, white chalky disks derived from coccolithophoridae and schizospheres. Swelling clays are absent or present in very low concentrations; illite is the principal clay mineral. With few exceptions calcite content in the marlstone unit exceeds 40%. Occasional thin dense limestone layers contain up to 85% calcite. Carbonate content in the shale units is typically less than 40% but ranges from 23% to 75% (in rare thin limestone beds). The limestones are layers rich in coccoliths and schizospheres which may be further enriched in carbonate by cementation.

As mentioned, bivalves are common in Unit II but are absent or rare in the other units. This appears to be the only consistent lithologic difference between the upper and middle shales. Although positive taxonomic identification was not made, the bivalves correspond to *Posidonia bronni* (now either *Bositra buchi* or *Steinmannia radiata*) and/or *Inoceramus* (now *Pseudomytiloides dubius*). Often the shells are concentrated in thin layers. Their distribution suggests that conditions during the deposition of Unit II were more tolerable to benthos, e.g. dysaerobic (Tab. 2; see photos in Littke et al., 1991b; Savrda and Bottjer, 1991), than those during deposition of Units I and III. Their rather irregular vertical occurrence within Unit II suggests that tolerable living conditions were episodic (see Seilacher, 1982) and irregularly distributed on the seafloor, such that areas of colonization continuously shifted to suitable but localized habitats.

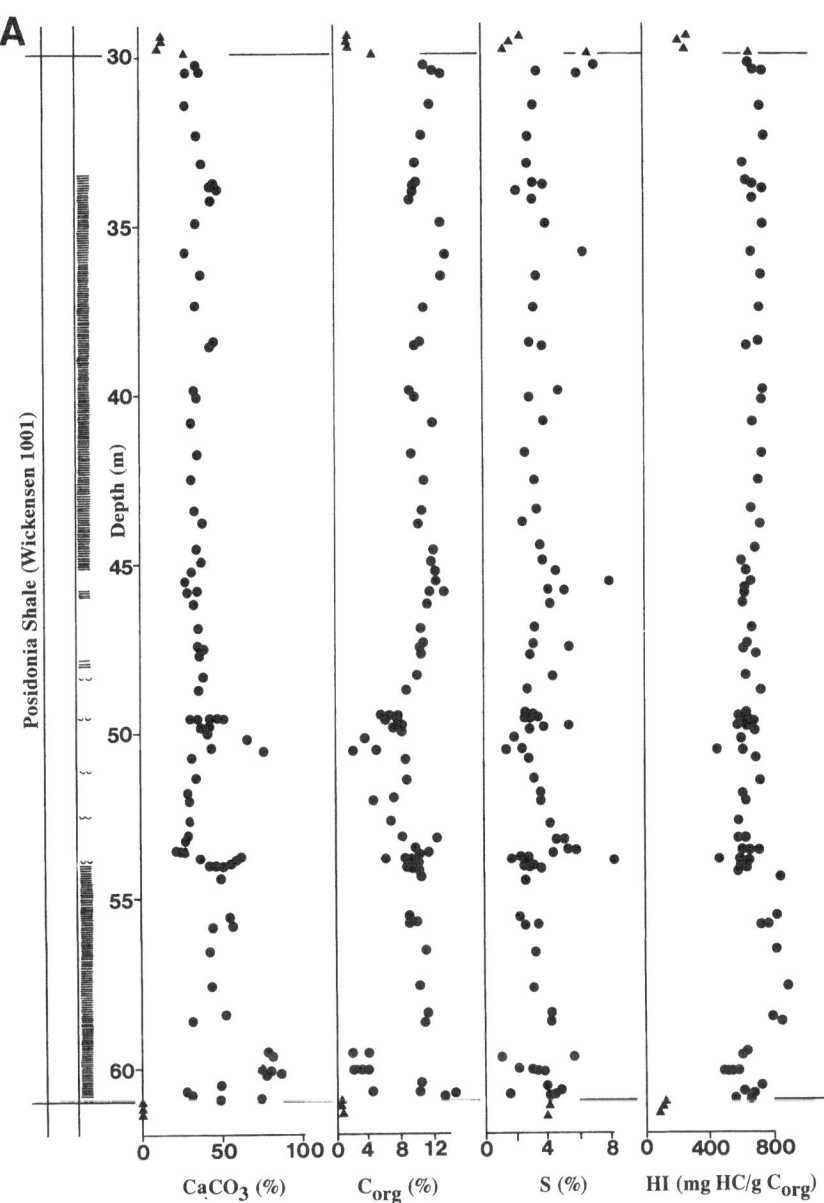

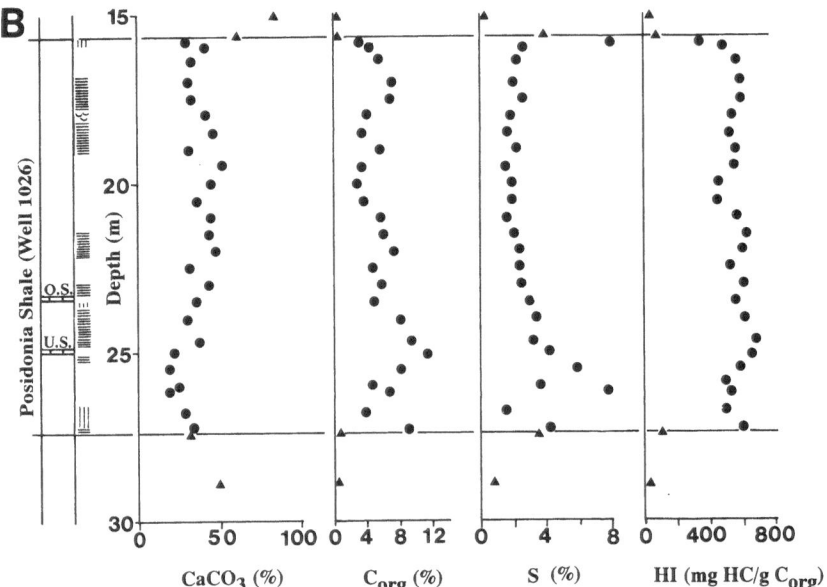

B Posidonia Shale (Well 1026)

Depth (m)

CaCO₃ (%)   C$_{org}$ (%)   S (%)   HI (mg HC/g C$_{org}$)

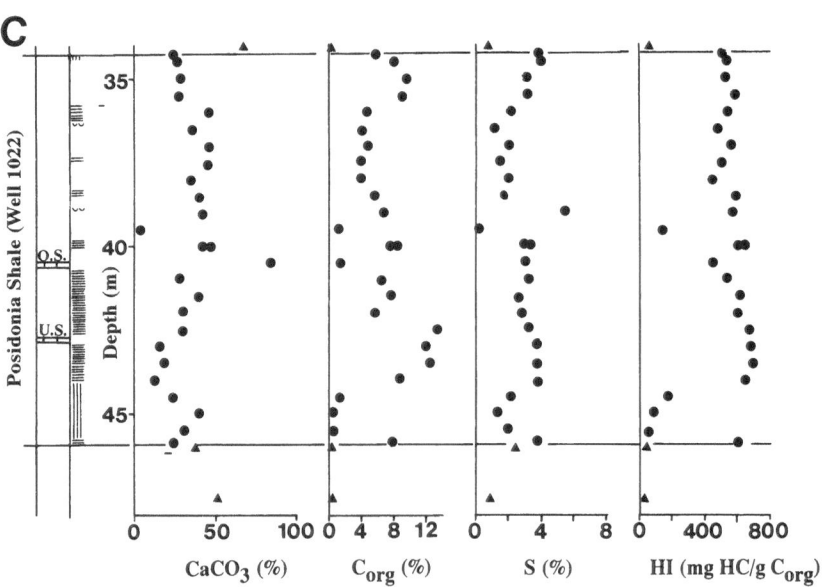

C Posidonia Shale (Well 1022)

Depth (m)

CaCO₃ (%)   C$_{org}$ (%)   S (%)   HI (mg HC/g C$_{org}$)

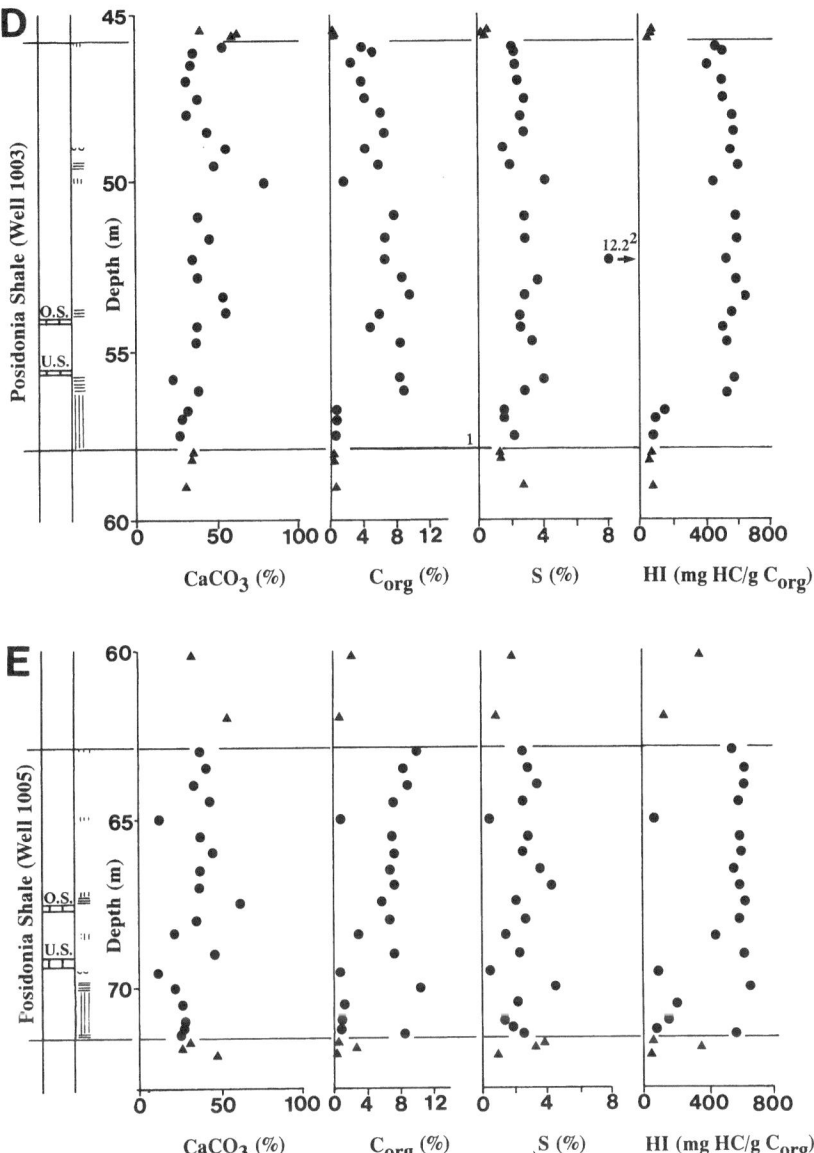

Fig. 25:   Lithologic log of the Wickensen core (A) and of wells 1026 (B), 1022 (C), 1003 (D), and 1005 (E) with principal geochemical values; $CaCO_3$, carbonate carbon recalcualted as wt % calcite; C_{org} wt % organic carbon; HI, hydrogen index (mg hydrocarbon-equivalents/g C_{org}); S, wt % total rock sulphur. Bioturbation (vertical bars), lamination (horizontal bars), and fossil hash layers are shown next to the left column. OS = "Oberer Stein" limestone, US = "Unterer Stein" limestone, [1] = basal C_{org}-rich layer ("Tafelfleins") not sampled, [2] = sample with pyritized macrofossil (from Littke et al., 1991a).

In southern Germany, the PS consists of fine-grained calcareous shales and marlstones with two, three or four interbedded, thin (<30cm) limestones which occur in the lower half of the PS. Two limestones ("Oberer Stein", "Unterer Stein") are used as stratigraphic marker horizons (Fig. 25B-E). The calcareous shales are mainly dark and finely laminated, but there are also several lighter coloured and bioturbated sections ("Fucoidenhorizonte"), especially near the top and bottom of the PS. Generally, a first, thin black shale at the base ("Tafelfleins") is overlain by gray, bioturbated shales ("Aschgraue Mergel"; Fig. 25B-E). Above the "Aschgraue Mergel", a thicker black shale follows which is partly laminated and contains several thin (<3cm) fossil hash layers and limestone beds (see above). The thickness of the PS varies from 8.6 to 12.1m in southern Germany and is slightly decreasing from south to north (Table 12; see Riegraf, 1985 for more details). In contrast, the thickness of the PS in the Wickensen core in northern Germany is 27.4m, some two-to-three times greater. As in both regions the PS covers the same three ammonite zones, i.e., *tenuicostatum, falciferum,* and *bifrons*, it can be regarded as (roughly) a time-stratigraphic unit. Thus, the thickness differences indicate that sedimentation rates of the PS were 2-3 times higher in northern Germany than in southern Germany.

Contact relationships are also somewhat different. In northern Germany, erosional disconformities mark the base and the top of the PS as described in detail by Littke et al. (1991b). Inbetween, there is only one fossil hash layer (coquina, about 2-3 cm thick) composed of fragments of thin shells, belemnites and phosphatic material (3.3 wt-% P), disseminated pyrite and a mud matrix which may represent a hiatus or an erosional event. In southern Germany, the basal contact of the PS with the underlying rocks appears to be abruptly transitional, i.e., nonerosional, whereas the upper contact of the PS is an abrupt erosional surface as in northern Germany (Loh et al., 1986; Rotzal, 1990; Bauer, 1991; Littke et al., 1991b).

4.4   Geochemistry and petrology of lithologic units in northern Germany

In northern Germany, carbonate, C_{org} and HI values of the under- and overlying sediments (Fig. 25A, Tab. 13) are significantly different from those of the PS. In these shales, the proportion of terrestrial to marine organic macerals is greater than in the PS (see Table 3 in Littke et al., 1991b). The lower and upper boundaries of the PS are geochemically and petrographically abrupt. There is no transition in these properties between the underlying Pliensbachian and the Lower Toarcian; C_{org}-rich samples appear suddenly and immediately above the basal contact. Only the deepest sample of the overlying sediments, located 20 cm above the PS contact, displays transitional properties, i.e., high carbonate, C_{org} and HI values (Fig. 25A). It seems likely that this narrow zone has simply incorporated carbonate and organic matter by reworking of the underlying Posidonia facies.

The PS displays considerable consistency in its bulk geochemical and petrographic parameters (Fig. 25A). Thus, the depositional milieu and production and preservation of organic matter must have generally been very uniform. However, some variation among the three units indicates small differences in depositional controls and processes.

Well	Thickness (m)	CaCO$_3$ (%)	C$_{org}$ (%)	S (%)
1026 (25)	11.7	34.6 ± 8.9	5.8 ± 2.2	3.0 ± 1.8
1022 (26)	11.6	33.9 ± 15.3	6.2 ± 3.6	2.9 ± 1.1
1003 (23)	12.1	40.6 ± 12.3	5.2 ± 2.7	3.0 ± 2.1
1005 (21)	8.6	30.9 ± 14.0	5.3 ± 3.5	2.5 ± 1.1
Wickensen	27.4[*]	42.9 ± 15.1	9.2 ± 2.9	3.6 ± 1.2

Table 12:   Thickness, carbonate- (CaCO$_3$), organic carbon- (C$_{org}$) and sulphur-percentages (average ± standard deviation) of PS in four wells from the Schwäbische Alb and in well Wickensen 1001 from northern Germany. In parentheses: number of analyzed samples. [*]Thickness corrected for dip.

Alginite B (with liptodetrinite), derived from small phytoplankton (<20 μm), is the most abundant maceral in all units. Bituminite is second in volumetric importance. Less abundant are large alginitic phytoplankton particles (alginite A) of *Tasmanales* and *Leiosphaeridales* types (Teichmüller and Ottenjann, 1977) and terrestrially derived vitrinite and inertinite. Alginite A (~ syn. telalginite) and alginite B (~ syn. lamalginite) were differentiated by the length of fluorescing particles in sections perpendicular to bedding only (see Hutton et al., 1980 for discussion). Almost all of the organic matter occurs as well defined discrete particles, i.e., there is little or no submicroscopic organic matter (see Fig. 3). All of the macerals appear to be uniformly disseminated within the rock matrix. Concentration of organic matter in thin layers or lamina was not noted.

Except for the difference in carbonate content, the marlstones of Unit I and calcareous shales of Unit III are geochemically identical, i.e., ranges and averages of C$_{org}$ and HI values are virtually identical (Table 13). However, there are several thin limestone beds in Unit I which have relatively low C$_{org}$ contents (average 3.4 wt%, Table 13). Similarly, the HI values of the limestones are lower (average 638) than those of the associated marlstones (average 735). This difference may be related to variations in maceral composition, because the only petrographically evaluated limestone contains a smaller proportion of alginite B and a higher proportion of alginite A than the marlstones.

Stratigraphic interval	N	Carbonate (wt %)		C_{org} (wt %)		HI (mg hc/g C_{org})	
		Range	Av.±St. Dev.	Range	Av.±St. Dev.	Range	Av.±St. Dev.
Aalenian Calc. Shale	9*	10.9-19.6	14.8±2.5	1.4-2.5	1.7±0.3	219-467	354±86
Lower Toarcian:							
III Calc. Shale	25	28.3-39.6	36.9±5.7	8.9-13.5	10.8±1.4	664-749	718±23
II Calc. Shale	14	22.9-38.8	31.8±5.6	10.0-13.1	11.1±1.0	660-718	684±16
Low C Zone	18	29.3-75.1	40.7±12.6	1.9-8.0	6.7±1.9	486-710	666±51
I Marlstone	19	30.0-61.6	48.2±9.1	8.7-14.7	10.3±1.5	712-803	735±33
Limestone	7+	74.6-85.0	79.7±3.3	2.1-4.6	3.4±1.0	584-702	638±43
Pliensbachian Shale	8	0.7-3.7	1.3±1.1	0.8-1.0	0.8±0.1	110-186	146±25

Table 13:   Summary of carbonate, organic carbon (C_{org}), and hydrogen index (HI) data of the Wickensen well, northern Germany (Littke et al., 1991b).
*Excludes transitional sediments at base.
+Excludes coquina layer at top of Unit I.

Two distinctive geochemical facies are recognized within Unit II (Table 13). One, referred to as calcareous shale, is similar in carbonate and C_{org} contents to the calcareous shale of Unit III, but has a slightly lower HI value (average 684 vs 718). The second facies, here referred to as the 'low-carbon-zone' (LCZ), has distinctly lower C_{org} contents (average 6.7 wt%) than either Units I or III or the associated calcareous shale facies of Unit II and, like the latter, has a lower HI value (average 666). All 18 samples of the LCZ-facies occur between 48.70 and 53.16 m depth (Fig. 25). The LCZ is 'sandwiched' within the Unit II calcareous shale facies.

Molecular hydrocarbon compound ratios calculated from extract data clearly differentiate between the PS and the underlying Pliensbachian and overlying Aalenian sediments (Table 14). This is true for more general parameters, like the Carbon Preference Index (CPI) of n-alkanes, as well as for biomarker (sterane and hopane) ratios. Within the PS, consistency of some parameters is contrasted by small differences in others.

The higher CPI values in the sediments over- and underlying the PS reflect their larger content of terrigenous components. Differences within the PS are not significant. Although the pristane/phytane ratio has been demonstrated to be of limited value as an indicator of levels of anoxia (ten Haven et al., 1987), the higher values in the Pliensbachian and Aalenian rocks probably reflect both the higher terrestrial organic matter contribution and more oxic depositional environments. The pristane/phytane ratios significantly exceeding unity indicate that the PS was not deposited under hypersaline conditions.

	Pliensbachian	Lower Toarcian				Aalenian
		Unit I	Unit II	Unit II	UnitIII	
	(Shale)	(Marlstone)	(LCZ)	(Calc. Shale)	(Calc. Shale)	(Calc. Shale)
Pristane/Phytane	2.71 (5)	1.76 (4)	1.45 (3)	1.76 (1)	1.38 (8)	3.28 (6)
CPI (C_{19}-C_{25})	1.14 (5)	1.05 (4)	1.02 (3)	1.01 (1)	1.01 (8)	1.13 (6)
Phytane/n-C_{18}	0.40 (5)	1.12 (4)	1.67 (3)	2.72 (1)	1.93 (8)	0.85 (6)
C_{27} neohopane/(C_{27} neohopane + C_{27} 17α (H)-hopane)	0.16 (1)	0.32 (4)	0.32 (3)	0.24 (1)	0.32 (14)	0.21 (4)
17α (H)-hopane/(17α (H)-hopane + moretane)	0.73 (1)	0.82 (4)	0.79 (3)	0.85 (1)	0.83 (14)	0.75 (4)
C_{29}/C_{27}-steranes	2.12 (1)	1.29 (4)	1.59 (3)	1.72 (1)	1.44 (14)	1.86 (4)
C_{29}/C_{28} steranes	2.41 (1)	2.00 (4)	1.43 (3)	2.50 (1)	1.97 (14)	2.21 (4)

Table 14: Average molecular compound ratios for the lithologic units in the Wickensen well, northern Germany (Littke et al., 1991b). Numbers in parentheses indidcate number of samples.

Absolute quantification of the hydrocarbons in the bitumen fractions (data not presented) shows that hydrocarbon generation in Unit I has proceeded farthest in the Wickensen well (0.53% R_o), i.e., highest yields of n-alkanes in the C_{15}-C_{20} range occur. This may be an effect of maceral composition, i.e., higher bituminite content in the marlstones which may be the most generative maceral at early stages of maturation. This could also account for the lower phytane/n-C_{18}-ratio displayed by the Unit I marlstones compared to Unit II and III. Further, as relatively high values of pristane/phytane ratios are known to be a consequence of early thermal hydrocarbon generation which favours pristane formation (Goossens et al., 1986), this phenomenon could also account for the slightly higher pristane/phytane ratio of Unit I compared to Unit II (LCZ) and Unit III. Variations in degree of hydrocarbon generation among the Posidonia members may obsure direct and simple palaeoenvironmental interpretations of molecular parameters.

The two selected hopane ratios (Table 14) distinguish between the over- and underlying shales on one hand and the PS on the other. At a given maturity level the 17α (H)-hopane/(17α(H)-hopane+moretane) ratio reflects the relative amount of terrigenous contribution (higher moretane input). Thus, as expected, the Pliensbachian and Aalenian rocks have the lower values, whereas in the PS the ratio is uniformly higher within experimental error limits. The C_{27}-neohopane/(C27-neohopane + C_{27} 17α (H)-hopane) ratio is also facies dependent although the reason (taxonomic or other) is not yet known. Its uniformity (Table 14) again indicates the uniform organic facies of the PS.

High C_{29} sterane proportions are often typical of a high terrigenous contribution, although this does not hold in a strict sense because certain aquatic biota apparently biosynthesize C_{29} sterols in high abundance (Volkmann et al., 1986). The measured values are in accordance with higher terrigenous organic matter supply to the over- and underlying sediments. Variations among the three lithologic units of the PS are significant at least between Units I and III on one hand and Unit II (LCZ) on the other.

4.5   Nature and origin of lamination

Regular parallel lamination is the most prominent sedimentologic feature of the PS. Laminae thickness usually ranges from less than 1 up to 10 mm. Light and dark layers alternate. The layers are consistently undisturbed and uniform, unscoured and without internal current structures, thus indicating the absence of both burrowing endobenthos and effective bottom currents.

In the Wickensen profile, the macroscopic character of laminae varies through the PS. In Unit I, laminae on average are about 1.5 mm thick, whereas laminae in the calcareous shale (Unit III) have an average thickness of about 3 mm. Lamination is present but often difficult to discern in Unit II. Lamination is absent in the uppermost 3.5m of Unit III. Light-dark laminae may represent a depositional pair, i.e., perhaps a periodic event. Assuming equivalent periodicity, the contrast in thickness indicates that the rate of sedimentation of the calcareous shale facies was about twice that of the marlstone. This distinction supports the view that deposition of the shale members was strongly influenced by a large supply of clay detritus, whereas terrigenous influx was considerably lower during the accumulation of the marlstone facies.

Geochemical analysis of eleven pairs of handpicked adjacent light and dark laminae (Fig. 26) reveals the following.

(i) Most light laminae are enriched in carbonate; absolute calcite enrichment is in the range 1-20% and averages about 9%; relative enrichment is in the range 3-66% and averages 30%. However, carbonate content in a light layer from an overall dark interval may be lower than in a dark layer taken from an overall lighter coloured interval. Hence the subdivision of light and dark layers is rather subjective. Exceptions to the above (Fig. 26) probably result from the difficulty of obtaining clean separation of light from dark layers and vice versa by the handpicking procedure.

(ii) C_{org} contents are consistently greater for dark than for light layers (Fig. 26; absolute enrichment 0.1-4.0%, average 1.5%; equivalent to average relative enrichment of 18%; see Weedon, 1986).

(iii) Consistent differences in HI values between light and dark layers are not evident. Most differences are within limits of analytical error.

(iv) Sulphur contents are often higher in dark than in adjacent light layers. Absolute enrichments range from 0.37 to 1.34%, equivalent to a relative difference of 10-35%. The sulphur-enriched dark laminae also show an approximately 20% relative enrichment of clay compared to light layers. Thus the local (layer by layer) preferential fixation of sulphide in dark (clay-rich) laminae is probably due to greater availability of clay-mineral iron. The observations support the view that iron availability is an important factor which controls sulphide concentration in the PS. The pair with appreciably greater sulphur in the light layer contains roughly equal levels of clay.

(v) Terrestrial macerals (vitrinite, inertinite, sporinite) are more abundant in dark than in light bands, while there is slightly (6-7%) more marine alginite in light layers.

(vi) Organic matter occurs as discrete, homogeneously distributed particles that are usually 10-200 μm in diameter and less than 20 μm thick in dark as well as in light laminae. These particles do not form lenses or layers as in many other oil shales (e.g., Messel Shale, Jankowski and Littke, 1986) where lamination may be due to seasonal deposition of plankton blooms. The disseminated aspect of the macerals does not support the view that widespread bacterial or algal mats were a characteristic feature of the PS depositional setting as proposed by Kauffman (1979).

There is no sedimentological or geochemical indication for differences in the level of anoxicity between the two types of laminae. On the contrary, the preservation of large amounts of hydrogen-rich alginite and especially bituminite (present in both light and dark layers) which are labile under oxic conditions indicates that the alternation of light and dark laminae did not result from variable preservation conditions. Also, the perfect parallel lamination contradicts a diagenetic origin for these laminae. Periodic changes in the proportion of marine (carbonate) to terrestrial (clay) input best account for the lamination. Assuming a constant flux of detrital clay, our data on laminae pairs indicate that the carbonate supply (i.e., plankton productivity) need only have increased by about 30% (on average) during the accumulation of light layers.

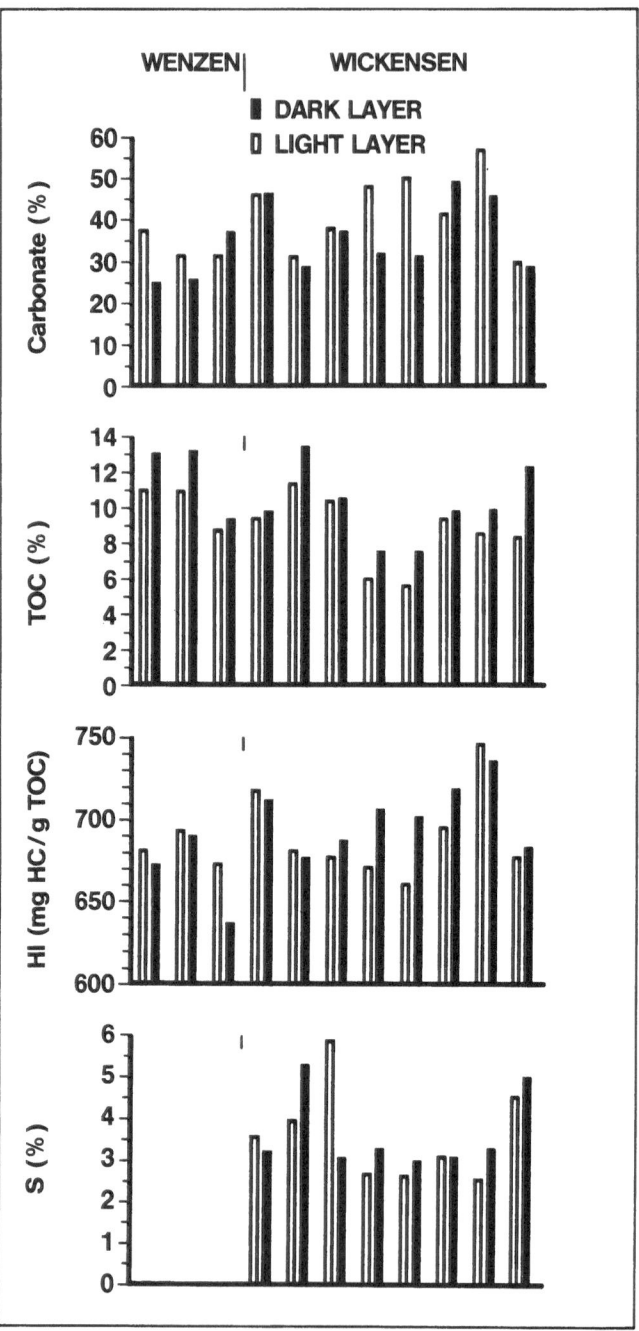

Fig. 26: Compositional observations of adjacent pairs of light and dark laminae from Wenzen (3 pairs; see Fig. 22) and Wickensen (8 pairs). No sulphur data were measured for the Wenzen samples (from Littke et al., 1991b).

In a nearby surface section, which is lithologically and stratigraphically similar to the Wickensen core, Loh et al. (1986) have recognized each of the three ammonite subdivisions of the Lower Toarcian, i.e., *tenuicostatum, falciferum* and *bifrons* zones. Hallam (1975, p. 23) estimates that the average time value for each Jurassic ammonite zone is one million years. Thus, it is surmised that the Wickensen PS section accumulated during roughly a 3 million year interval. Using this time duration, the calculated sedimentation rate (compacted) for the PS is about 9 mm/1000 years. This rate is similar to the range of values, 1-24 mm/1000 years, determined by Schwab (1976) for several ancient cratonic basins, i.e., basins with low sedimentation rate. Based on the laminae thickness averages, we estimate that 5000-6000 laminae pairs comprise the Wickensen core. Calculations suggest that each pair represents a 500-600 year interval. Clearly, a seasonal period for deposition of the light-dark pairs seems excluded. This also holds true if a shorter duration for PS deposition as proposed by Jenkyns (1985, 0.5 million years) is assumed (see discussion in Chapter 4.8).

4.6   Geochemistry and petrology of PS in southern Germany

Basic geological and geochemical data are summarized in Table 12 and Fig. 25. Three profiles (1026, 1022, and 1003) are similar with respect to thickness and average C_{org} and sulphur contents. Furthermore, C_{org} and sulphur data show similar depth trends (Fig. 25B-D). Generally, there is an abrupt increase in C_{org} at the base of the PS, e.g., values below 0.5% occur directly below and values greater than 8% directly above the base in well 1026. This basal black shale ("Tafelfleins"; not sampled in well 1003) is overlain by thin bioturbated horizons ("Aschgraue Mergel", Fig. 23) with lower C_{org} and lower sulphur contents. In the interval above ("Unterer Schiefer", Fig. 23) C_{org} and sulphur contents increase to high (often maximum) values below the "Unterer Stein" limestone. Maximum C_{org}-values in this section exceed 10%. From the "Unterer Stein" limestone to the top of the PS, C_{org} and sulphur values in cores 1026, 1022, and 1003 gradually but irregularly decrease to about 4%, although there are a few exceptions, e.g., samples with C_{org} above 6% near the top of the PS in well 1026. The uppermost samples may be extremely enriched in sulphur, e.g., 8% in well 1026. Marlstones overlying the PS contain less than 0.5% C_{org}. Carbonate contents generally increase upwards from about 30% near the base of the PS to greater than 40% three metres below the top; above this level carbonate contents decrease towards the top.

The major difference between these three profiles from the southern and central part of the Schwäbische Alb and the northern profile (well 1005; Figure 25E) is its smaller thickness and the trend of upwards increasing (rather than decreasing) C_{org} and sulphur contents between "Unterer Stein" and the top of the PS.

The PS in the Wickensen core from northern Germany is thicker and richer in C_{org}, carbonate and sulphur (Table 12) than its counterpart from southern Germany. The abrupt increase in C_{org} at the bottom and sharp decrease at the very top of the PS across the upper contact (Fig. 25A) are similar to those in the Schwäbische Alb. The Wickensen profile is overall more homogeneous with respect to bulk geochemical composition than the PS in the Schwäbische Alb area, e.g., there are no C_{org}-lean intervals or sections in which low HI-values prevail.

By means of optical microscopy, the same four groups of organic particles were found as in northern Germany (Table 15). Average percentages of these four groups are similar in all profiles, e.g., the bulk of the organic matter consists in all cases of alginite B and liptodetrinite. Interestingly, terrestrial-derived macerals are - relative to marine macerals - more abundant in samples from wells 1003 and 1005 from the northern and central part of the Schwäbische Alb than in those from the southern locations and well Wickensen. This may indicate a closer proximity to the source areas for terrigenous clastic debris or a stronger degradation of marine organic matter in the northern part of the Schwäbische Alb. Also of interest is the low average percentage of bituminite in samples from well 1005. This maceral is almost exclusively found in liptinite-rich black shales deposited under conditions of extremely good preservation of organic matter (e.g., Stein et al., 1988; Teichmüller, pers. communication). Its scarcity in the PS of well 1005 indicates diminished organic matter preservation compared to the other locations.

Well	Vitr.+Inert. (Vol.-% of TM)	Bituminite (Vol.-% of TM)	Alginite (Vol.-% of TM)	Alginite B + Liptodetr. (Vol.-% of TM)
1026 (13)	3.4 ± 3.1	9.1 ± 6.9	7.8 ± 5.4	79.7 ± 9.6
1022 (11)	2.5 ± 4.3	4.1 ± 3.2	6.0 ± 5.1	86.1 ± 10.3
1003 (11)	8.0 ± 16.1	6.2 ± 5.0	6.4 ± 7.8	79.7 ± 16.5
1005 (7)	7.0 ± 9.4	1.3 ± 1.4	7.7 ± 6.3	84.1 ± 9.6
Wickensen (32)	4.1 ± 2.0	18.0 ± 8.3	3.0 ± 3.1	75.0 ± 9.2

Table 15: Average volume percentages (of total macerals; average ± standard deviation) of vitrinite and inertinite, bituminite, alginite A, and alginite B and liptodetrinite of PS in four wells from the Schäbische Alb and in well Wickensen 1001 from northern Germany. In parentheses: number of analyzed samples.

Another measure for the preservation of organic matter is the hydrogen index (HI) value. High HI values (>400) indicate preservation of hydrogen-rich (usually oil-prone) organic matter. Low HI values are either due to input of primarily hydrogen-poor terrigenous organic matter, pre- and syndepositional organic matter degradation, or advanced maturity (hydrocarbon generation and expulsion). HI-values of the PS from the Schwäbische Alb are significantly lower than those of the PS from well Wickensen 1001 (Table 16). In view of the similarity in maceral composition and maturity, this is interpreted to reflect more favourable conditions for preservation of sedimentary organic matter in northern Germany during the Lower Toarcian. Nevertheless, compared to most other organic matter-rich sediments, average HI values of the PS in southern Germany are still high and indicate oxygen-deficient bottom water during most of the time of deposition. If HI values are taken as a measure of organic matter preservation, organic matter in the PS of the southern and central parts of the Schwäbische Alb (wells 1026, 1022, and 1003) is on an average less degraded than in the northern part (well 1005), where the average HI value is more than 10% lower. This is due to the interbedded occurrence of several bioturbated C_{org}-poor, clay-rich, carbonate-poor intervals (Fig. 25E) in which hydrogen-poor organic matter prevails.

Well	PI $(S_1/(S_1+S_2))$	HI (mg hc/g C_{org})	T_{max} (°C)
1026 (25)	0.03 ± 0.01	548 ± 74	426 ± 2
1022 (26)	0.03 ± 0.01	504 ± 182	426 ± 5
1003 (23)	0.10 ± 0.03	485 ± 160	433 ± 7
1005 (21)	0.09 ± 0.06	437 ± 226	431 ± 3
Wickensen (83)	0.05 ± 0.03	698 ± 70	426 ± 3

Table 16:   Rock Eval pyrolysis data (average ± standard deviation) of PS in four wells from the Schwäbische Alb and in well Wickensen 1001 from northern Germany. In parentheses: number of analyzed samples. PI = production index, HI = hydrogen index. T_{max} = temperature of maximum pyrolysis yield.

Information on the maturity of organic matter is derived from the temperature of maximum pyrolysis yield (T_{max}) and the so-called production index (PI = S1/(S1+S2)) listed in Table 16. Both parameters generally increase with increasing degree of thermal kerogen conversion and hydrocarbon generation (Espitalié et al., 1977). Average T_{max} and PI values of the Wickensen core are in accordance with the measured maturity of 0.53% vitrinite reflectance (Littke et al., 1991b). Based on T_{max} and PI values about the same maturity level was established for samples from wells 1026 and 1022 (Table 16), where the PS is regarded as immature to marginally mature petroleum source rock, i.e., no or only marginal petroleum generation occurred during its burial history. The PS in wells 1003 and 1005 is slightly more mature as indicated by higher average T_{max} values and higher average PI values. Peak oil generation was, however, not reached in the PS at these sites as obvious from a comparison with geochemical data (e.g., T_{max} values) of more mature PS cores (Rullkötter et al., 1988a), i.e., vitrinite reflectance is not expected to exceed 0.6%. For Jurassic sediments drilled in an adjacent well (Urach 3), vitrinite reflectance values of about 0.5% are reported (Buntebarth et al., 1979).

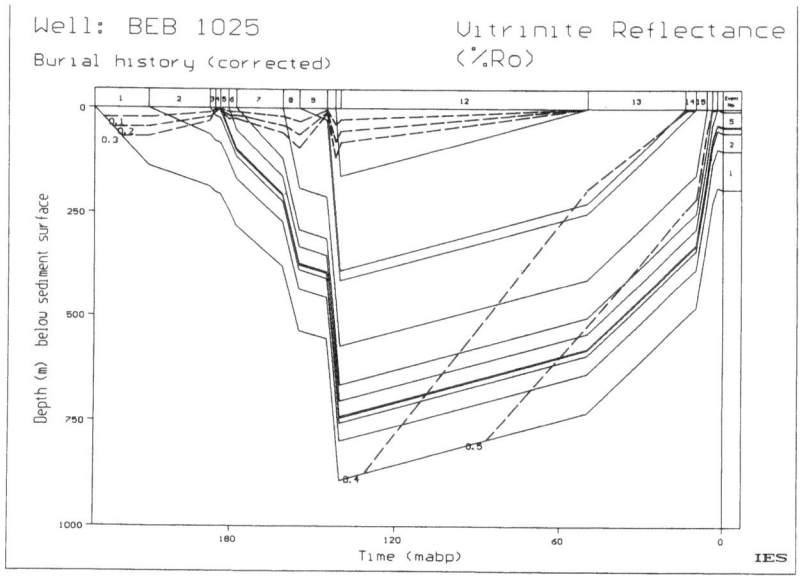

Fig. 27:   Burial history of the PS (black) in the Schwäbische Alb (well 1025) and evolution of vitrinite reflectance (from Bauer, 1991).

The maturity level of the PS in the Schwäbische Alb area, especially at the northern and central location, is rather high in view of the low maximum burial depth of this rock. This "high" maturity is confirmed by biomarker studies (Bauer, 1991). The observation of a "too high" maturity level is not adequately explained even if high heat flows (>100 mW/m^2) are assumed to have persisted during the entire Tertiary as revealed by numerical simulations of temperature histories (see Chapter 5.6 for more details of the temperature history modelling). It can only be explained if higher sedimentation and erosion rates than previously assumed and/or high heat flows existed during the Cretaceous. A model of the burial history and vitrinite reflectance evolution for the southern part of the Schwäbische Alb is shown in Fig. 27 (see Bauer, 1991 for more details). It is based on the assumption that the total thickness of the Upper Malmian in the study area (well 1025, close to well 1026, see Fig. 23) was 400m, and that no Cretaceous sediments were deposited (Geyer and Gwinner, 1962). Furthermore, high heat flow values were assumed for the Upper Malmian (90 mW/m^2) and for the Cretaceous and Tertiary (100 mW/m^2) with exceptionally high values for times of volcanic activity in the Neogene (Urach and Hegau volcanism). Only with the assumption of this high heat flow, a fit between calculated and measured maturity parameters (Fig. 28) is achieved. Lower heat flows can only be assumed, if the thickness of eroded sediments was greater, i.e., if an erosion of a thicker Upper Malmian section or of Cretaceous sediments is assumed.

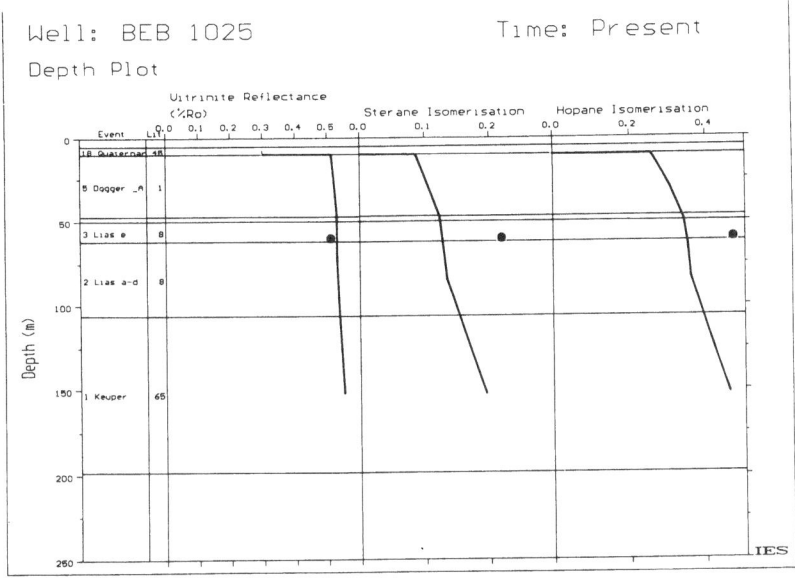

Fig. 28:   Calculation of maturity parameters based on burial and temperature history and comparison with measured values for the PS (from Bauer, 1991).

4.7   Sulphate reduction and primary sediment composition

In the PS pyrite occurs macroscopically as replacement of carbonate shells and as irregular nodular masses; belemnites are consistently unpyritized. Framboidal pyrite is the most prominent microscopic variety. In addition isolated euhedral grains, typically 1-3 μm in diameter, are ubiquitous. Total sulphur contents range from about 1 to 8 wt% (Fig. 25). Microscopic estimates of volume per cent pyrite compared to the elemental sulphur analyses and sulphur-iron ratios (Rotzal, 1990) indicate that the bulk of the sulphur is bound in the various forms of pyrite.

Differences in the ranges and average values of total sulphur content of the three PS units in northern Germany are not particularly large. The marlstones and limestones of Unit I have a slightly lower sulphur content than the calcareous shales of Units II and III. The average absolute sulphur content of the shales from the LCZ of Unit II (average 3.18%) is lower than that of the more organic matter-rich shales from Unit II and III (Littke et al., 1991b). Two separate trends are evident from the C_{org} vs sulphur (C/S) plot (Fig. 29A+B). The marlstone and shale samples of Unit I and III define a trend with a slope (S/C) of about 0.39 and intercept at - 0.74% sulphur, whereas Unit II samples (both calcareous shale and LCZ-facies) display a trend with S/C slope of 0.34 and sulphur axis intercept of 0.87%. The trend line for PS from southern Germany is similar to the trend line for Unit II samples from northern Germany in terms of intercept with the sulphur axis and slope (Fig. 29B). The C/S plot for Black Sea sediments (Fig. 29A) intersects the sulphur axis at 1-2%. Following the reasoning of Leventhal (1983), the absence of a significant positive sulphur intercept indicates that 'excess' sulphide added by water-column iron sulphide formation did not precipitate during PS deposition. In contrast to the Black Sea conditions, the observations suggest that an extensive H_2S-rich 'euxinic' water-column (several hundred metres) did not exist above the depositional interface of the PS.

When viewed collectively (i.e., as a single population) the PS data could be regarded as an extension of the 'normal (oxic) marine' trend (Fig. 29A) to higher C_{org} values. Recognizing that substantial bacterial sulphate reduction occurred within the Posidonia sediments at or near the sea-floor, it is quite likely that the bottom waters were generally anoxic. However, the comparisons above indicate that high H_2S-levels did not extend far above the depositional interface and that, whether permanent or ephemeral, the euxinic zone was relatively thin.

It is difficult to provide an unequivocal explanation which accounts for the distinctive trends displayed on the C/S plot (Fig. 29A, 29C). This is because the controlling variables of sulphide generation and fixation are many and constraints are few (see discussion by Fisher and Hudson, 1987: 70-71). The most important general constraint is that the two trends in Fig. 29A correspond to separate stratigraphic units. This indicates that the differences are due to primary (depositional and early diagenetic) parameters.

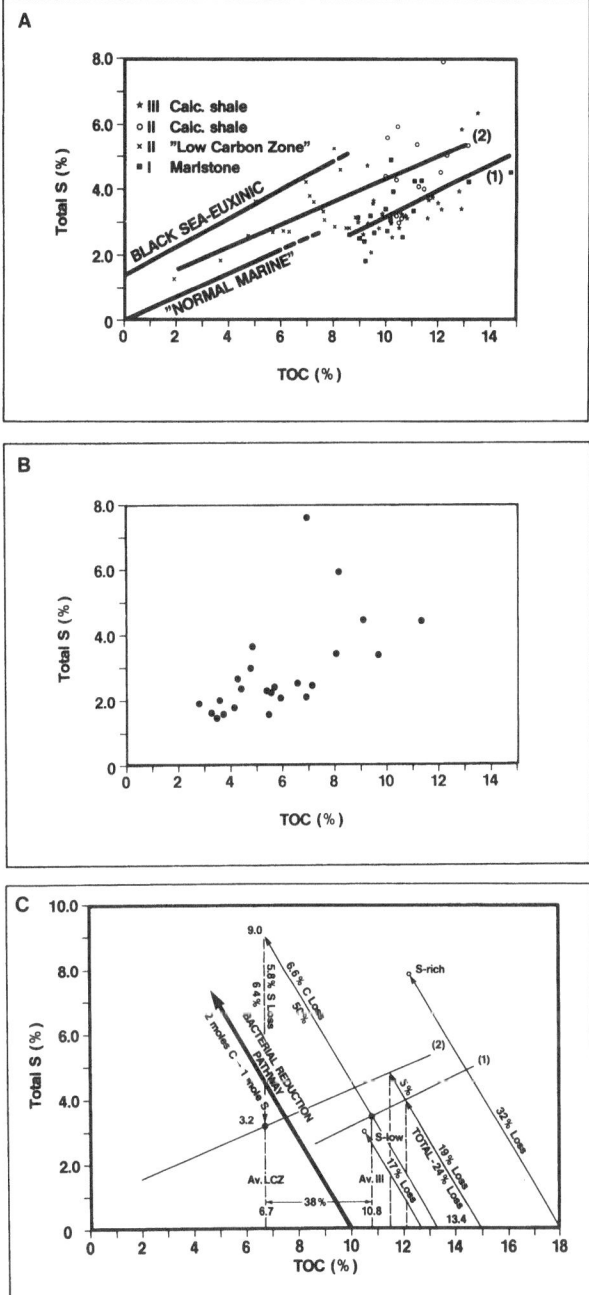

Fig. 29:   Weight -% plot of total sulphur (S) versus total organic carbon (TOC) for samples
from different units in the PS from well Wickensen (A) and for samples from well
1026 in southern Germany (B). Fig. 29C shows the bacterial sulphate reduction
pathway and - based on this - reconstructions of "original" organic carbon-values in
the PS (regression lines (1) and (2) are from Fig. 29A; see text for more explanation).
Regression line 1 is valid for Unit I and Unit III: S = 0.39 TOC -0.74; r = 0.79, N =
43; regression line 2 is valid for Unit II: S = 0.34 TOC + 0.87; r = 0.84, N = 32.

The positive relationship between C_{org} and sulphur contents displayed by the trends shows that H_2S generation (and hence sulphide fixation) is dependent of the amount of organic matter despite the rather high C_{org} contents of all the samples. This may be because only a part of the primary organic matter is reactive, i.e., metabolizable by sulphate reducers. Another explanation for the positive correlation is that both sulphur and C_{org} are controlled by terrigenous influx, i.e., sulphur contents are related to the amount of iron fixed in clay minerals and C_{org} is controlled by terrigenous nutrients. The latter assumption is supported by the positive correlation between sulphur and clay for the PS in well Wickensen (Littke et al., 1991b: Fig. 8).

Perhaps the simplest explanation for the differences is that bacterial sulphate reduction proceeded more extensively in the sediments of Unit II than in Units I and III during early diagenesis. At that stage, sulphate-reducing bacteria produced hydrogen sulphide utilizing organic matter (OM) as a substrate (equation 1). Hydrogen sulphide reacted with detrital iron to form iron monosulphide (equation 2) and iron disulphide (pyrite, equation 3; Berner, 1984; Leventhal, 1983).

$$\text{Eq. 1} \qquad SO_4^{2-} + 2CH_2O \qquad \rightarrow \qquad 2HCO_3^- + H_2S$$

$$\text{Eq. 2} \qquad 3H_2S + 2FeOOH \qquad \rightarrow \qquad 2FeS + S° + 4H_2O$$

$$\text{Eq. 3} \qquad FeS + S° \qquad \rightarrow \qquad FeS_2$$

According to equation 1, two moles of C_{org} are consumed to produce one mole of reduced sulphur. The compositional pathway for this process is shown on Fig. 29C. The projection of the alteration pathway from a specific sample point gives an approximation of the C_{org} content of the original sediment. This estimate is a minimum because it assumes that 100% of the H_2S generated was fixed as pyrite and also that there was no subsequent loss of carbon due to methanogenesis. As shown by the construction, bacterial alteration to the I-III trend line (line 1, Fig. 29C) results in a 19% loss relative to original carbon content. The position of the sample points and trend-line for Unit II are distributed farther along the bacterial alteration pathway. Specifically, the gap between the trend lines corresponds to a further carbon depletion of about 5%. For Unit II samples with highest sulphur contents, calculated carbon depletion is up to 32%, for those with lowest sulphur, carbon depletion amounts to only 17% (see constructions in Fig. 29C). The data analysis supports the proposition that the C/S distinction between Unit II and Units I and III may be explained primarily by more extensive bacterial alteration (greater H_2S generation and carbon loss) within the sediments of Unit II. Thus, the sulphur values (Table 12) can be used to recalculate the C_{org} content of the sediment prior to the onset of sulphate reduction (referred to as original C_{org} in the following text) according to:

$$\text{Eq. 4} \qquad C_{org}^* = C_{org} + 2S \cdot M_C/M_S$$

In equation 4, C_{org}^* is the original C_{org} weight percentage and M_C and M_S are molar masses of carbon and sulphur.

The calculated C_{org}^* (Table 17) are minimum values if, 1) part of the generated H_2S was lost from the sediment, i.e., not fixed in pyrite, 2) activity of methanogenic bacteria below the zone of sulphate reduction destroyed OM, or 3) thermogenic hydrocarbon generation and migration occurred. Geochemical data do not indicate that the second and third process - although being certainly of some importance - did significantly add to the diagenetic C_{org} loss, i.e., biomarkers indicative of methanogenic bacteria are not particularly abundant and thermal maturity is low.

Well	C_{org}^* %	OM^* %	SIL %	ARSED (g/(m²a))	ARCAR (g/(m²a))	ARSIL (g/(m²a))	ARC_{org}^* (g/(m²a))
1026	8.1	10.3	55.1	9.8	3.4	5.4	0.8
1022	8.4	10.7	55.4	9.7	3.3	5.4	0.8
1003	7.5	9.6	49.8	10.2	4.1	5.1	0.8
1005	7.2	9.2	59.9	7.2	2.2	4.3	0.5
Wickensen	11.9	15.3	41.8	23.0	9.9	9.6	2.7

Table 17: Weight percentages of organic carbon (C_{org}^*), organic matter (OM^*) and silicate (SIL) before sulphate reduction according to equations 4-6 in four wells from the Schwäbische Alb and in well Wickensen 1001 from northern Germany, and accumulation rates for bulk sediment (ARSED), carbonate (ARCAR), silicate (ARSIL, see equation 6), and organic carbon (ARC_{org}^*). All values are mean values for PS. Accumulation rates are calculated using the assumption of a continuous sedimentation over 2.5 million years.

Thus, the principal additional cause for organic carbon loss is assumed to be the first process, viz the loss of H_2S into the overlying water, possibly followed by sulphide oxidation to sulphate which in turn can be used by sulphate-reducing bacteria according to equation 1. Such a recurrent sulphur cycle is, however, limited in the PS by the fine-grained character and low permeability of the sediments and by the iron availability (Rotzal, 1990) which was - due to the high clay content - probably sufficient to fix most reduced sulphur as pyrite. We therefore assume that the C_{org}^* values presented in Table 17 do not much underestimate the C_{org} content of the original sediment (probably by less than 20% of the value).

Based on C_{org}^*-values the original OM can be calculated as

Eq. 5 $$OM^* = C_{org}^* \cdot 100/C_{OM}$$

In this equation OM^* is the OM weight percentage prior to the onset of sulphate reduction and C_{OM} is the carbon content of OM. In intrapolation of published data on carbon content of OM from the PS at different maturities (Rullkötter et al., 1988a), a numerical value of 78% is assumed for C_{OM}. OM^* values are summarized in Table 17.

Based on the above calculations, each PS sample can be regarded as primarily composed of three major components only: silicate, carbonate, and OM. The silicate weight percentage (SIL) is calculated by difference (equation 6; Table 17) and includes 1-2% of phosphatic material (mean P = 0.1%; Bauer, 1991).

$$\text{Eq. 6} \qquad SIL = 100 - CAR - OM^*$$

Controls on the bulk sedimentary composition of the PS prior to sulphate reduction are deduced from a triangular diagram (Fig. 30) in which original OM, carbonate, and silicate concentrations are plotted for wells 1026, 1005, and Wickensen. Data points of wells 1022 and 1003 fall into the same area as those of well 1026 and are therefore not shown. Samples from well 1026 (closed circles) define a trend line (1) showing that with increasing silicate, OM is also increasing. Thus, a (theoretical) carbonate-free sample would plot at 13% original OM, whereas a (theoretical) pure carbonate sample would contain no OM. The same general trend was also found for Wickensen samples (open circles; trend line 2), but a (theoretical) carbonate-free sample would be richer in OM there (about 20%). Both indicate that carbonate is a diluent to OM and that the organic biomass of precursors of coccoliths and schizospheres did not significantly contribute to the OM in the PS. In Wickensen, kerogen concentrations are at a given silicate/carbonate ratio consistently greater than in well 1026. In view of the similar silicate/carbonate ratios at both sites, this is explained as an effect of better preservation conditions for OM at the northern German site during PS deposition.

Most samples from PS of well 1005 (triangles) either plot between trend lines 1 and 2 or fall into the "silicate corner". The latter samples may represent times in which bioproduction was much reduced, therefore OM and carbonate are depleted. In addition, organic carbon loss was supported by oxidation within the sediments. The oxygen supply into the sediments is clearly documented by macroscopically visible bioturbation (Fig. 25E). The samples which plot above the trend line for PS from well 1026 are tentatively explained by a greater ratio of the flux of organic biomass over the flux of detrital silicate. Another possible explanation is better preservation of the samples of well 1005 compared to those of well 1026, but this is not supported by higher HI values.

Although other studies indicate that aerobic alteration, as in the water column, significantly diminishes HI values (Harvey et al., 1986 and references therein), greater intensity of anaerobic bacterial alteration and resulting carbon loss may account for the lower HI values of PS in southern Germany and in Unit II. A plot of carbon loss (again due only to sulphate reducers) vs HI (Fig. 31) for all PS samples from the Wickensen well (line 1) shows a decrease in HI with increase in carbon loss. This is also evident for samples exclusive to Unit II (line 2). The relationship seems to indicate that sulphate-reducing bacteria selectively destroyed a more hydrogen-rich component of the primary organic matter mixture.

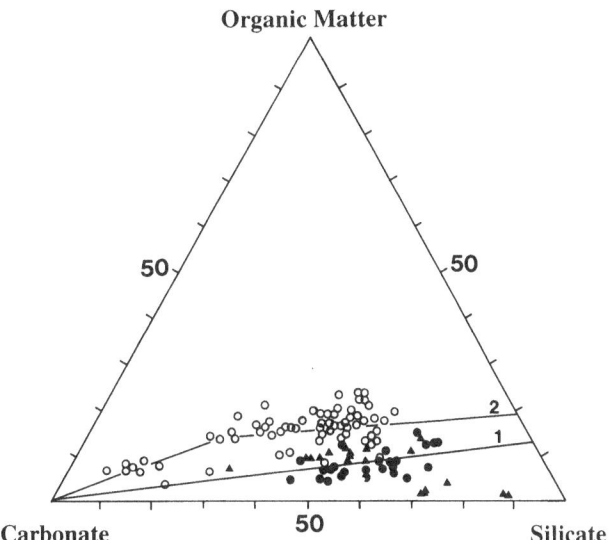

Fig. 30:   Triangular plot of original sediment composition (organic matter, silicate, carbonate) calculated according to equations 4-6 for PS from well Wickensen (open circles, see Littke et al., 1991b for more details), well 1026 (circles) and well 1005 (triangles). The composition of PS from wells 1022 and 1003 is similar to the one of PS from well 1026 (from Littke et al., 1991a).

Bacterial alteration and resultant carbon depletion does not, however, fully account for the low carbon level of the LCZ of Unit II. The average of the C_{org} content of the LCZ (6.7%) is about 38% less than the average (10.8%) of shales from Unit II. As reviewed before, relative carbon loss for Unit II samples in addition to that experienced by Units I and III averages about 5% with a maximum of less than 15%. As illustrated in Fig. 29C, a 38% carbon reduction by bacterial action would require an original C_{org} value nearly twice (13.4%) the LCZ average, i.e., a total carbon loss of about 50%, and a loss (i.e., lack of pyrite precipitation) of some 64% of the generated H_2S. As noted previously, other evidence indicates that more aerated bottom waters, perhaps associated with weak currents, may have characterized the depositional realm of Unit II; therefore, we suggest that water column oxidation and lateral transport of organic matter to deeper water Posidonia facies may be additional causes for the diminished accumulation of primary organic matter within the LCZ of Unit II. To what extent aerobic degradation (Harvey et al., 1986) on one hand and anaerobic degradation (Fig. 29) on the other hand caused the obvious hydrogen depletion of organic matter in the LCZ cannot be resolved.

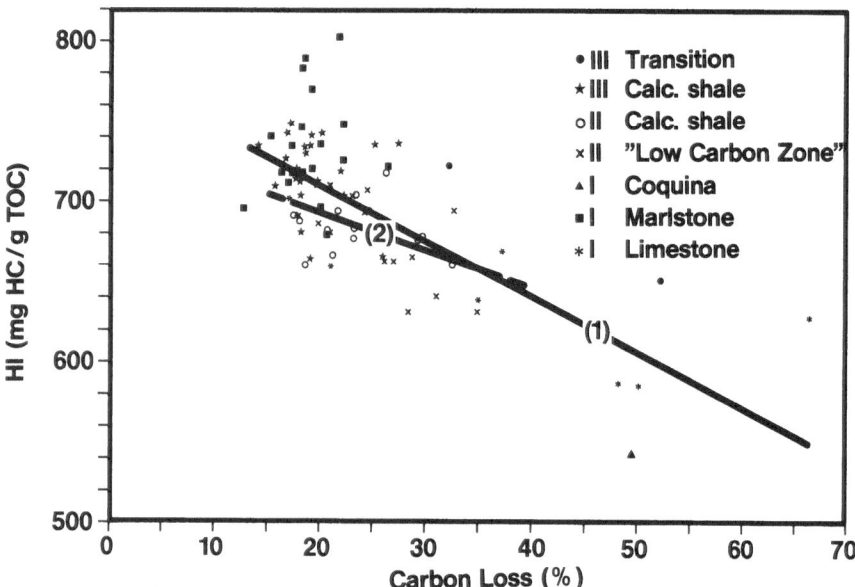

Fig. 31:   Plot of hydrogen index (HI) versus carbon loss (relative to original organic carbon), calculated according to equation 4 for samples from well Wickensen. Line 1 for all samples: HI = -3.2 C_{loss} + 775; r = -0.82, N = 84; line 2 for Unit II: HI = -2.3 C_{loss} + 738; r = -0.70, N = 31 (from Littke et al., 1991b).

4.8   Accumulation rates, primary productivity, and duration

Calculation of mass accumulation rates provides a means for comparison of lithified, and non-lithified, i.e., old and recent sediments. Furthermore, accumulation rates for individual components are - in contrast to weight percentages - independent of dilution effects by other components. The accumulation rate for bulk sediment is calculated according to equation 7.

Eq. 7   ARSED = LSR · (WBD - 1.026 Po/100)

where

ARSED   = mass accumulation rate (g/(cm^2·a))

LSR   = linear sedimentation rate (cm/a)

WBD   = wet bulk density (g/cm^3)

Po   = porosity (%)

Numerically, the term (WBD - 1.026 Po/100) is about 2.1 g/cm^3 for immature PS (Mann, 1987). The LSR is calculated dividing the thickness of the PS profiles (Table 12) by the assumed duration of the event. A duration of 2.5 million years was used for calculation of the values summarized in Table 17. For calculation of mass accumulation rates of the three major sedimentary components, percentages of carbonate, silicate, and C_{org} in the original sediment (Tables 12 and 17) were multiplied by ARSED and divided by 100. C_{org} rather than organic matter (OM) was selected to allow comparison with published data (Table 18).

Area	C_{org} %	HI (mghc/ g C_{org})	ARSED (g/(m^2·a))	ARC$_{org}$ (g/(m^2·a))	PP (gC/(m^2·a))
Site 645 (Baffin Bay)	1.1	85	60-200	0.9	50-80
Site 658 (NW-Africa)	2.0	261	120	2.0	200-320
Site 723 (Oman margin)	3.1	340	165	5.1	n.d.
Site 679 (Peru margin)	4.0	455	65	2.6	n.d.

Table 18:   Organic carbon percentages, hydrogen index (HI) values, bulk sediment and organic carbon accumulation rates, and palaeoproductivity in sediments of Baffin Bay (west of Greenland; Miocene to recent), on the continental margin offshore Northwest Africa (Pliocene to recent; Stein and Littke, 1990), offshore Oman (Pleistocene to recent; Prell, Nllisuma et al., 1989), and offshore Peru (Quaternary, ten Haven et al., 1990).

The mass accumulation rates for bulk sediment, silicate, and C_{org} are almost identical for PS from wells 1026, 1022, and 1003 from the southern and central part of the Schwäbische Alb (Table 17). Carbonate accumulation rates are slightly higher for PS from well 1003 than for PS from wells 1026 and 1022. Clearly, accumulation rates are lower in well 1005 for bulk sediment and silicate, and even more drastically for C_{org} and carbonate.

At Wickensen in northern Germany, a five times greater mass of carbonate and C_{org} than at Site 1005 was deposited during the Lower Toarcian, and three times more than at Sites 1003, 1022 and 1026. Silicate accumulation is about twice as high as in the Schwäbische Alb area. The factor by which accumulation rates are higher in northern than in southern Germany is greater for C_{org} than for carbonate. This supports that the greater organic matter percentage (Fig. 30) in PS of Wickensen compared to PS in the Schwäbische Alb area is indeed an effect of better preservation of organic matter and not or not only of greater productivity. Greater productivity should lead to an increase of both planktonic carbonate and planktonic organic matter content. This reasoning supports the previous conclusion of better preservation of organic matter in northern Germany which was based on the greater HI values.

Accumulation rates for organic carbon were determined for many subrecent sediments and are known to be related to primary productivity (Tab. 18), sedimentation rate, and water depth. In oxic depositional environments, organic carbon accumulation rates are often positively correlated with primary productivity and with sedimentation rate and negatively correlated with water depth. The relevant destruction rates in the water and on the sea floor are discussed in more detail by Stein (1991) based on Berger et al. (1989). For environments with anoxic bottom waters this relation is less clear. Due to the inhibited aerobic bacterial alteration in these settings, destruction of organic matter is slower. High sedimentation rates lead to a rapid burial of organic matter; this has a positive effect on organic carbon values in oxic seas. In anoxic seas, high sedimentation rates have a dilutant effect and lead to lower organic carbon percentages.

A comparison of Tables 17, 18, and 19 reveals for the PS, that a high primary productivity as typical for upwelling areas cannot be assumed. Areas of medium to high marine productivity such as Baffin Bay (between Greenland and Canada) and the continental margins off Northwest Africa, Oman, and Peru are characterized by much higher bulk sediment accumulation rates and lower C_{org} percentages (Table 18) than the epeiric Lower Toarcian sea (Table 17). Furthermore hydrogen indices are lower in sediments deposited under highly productive surface waters (Table 18) than in PS (Table 16). It is not yet clear, whether this phenomenon only mirrors a greater OM degradation in the water column or whether it is also an effect of greater OM degradation at the sedimentary surface and within the sediments due to greater availability of oxidizing agents. High bioproductivity can only be assumed for the Lower Toarcian, if the PS was deposited at much higher sedimentation rates and represents only a much shorter period (<<500.000 a) within the Lower Toarcian as previously proposed (Jenkyns, 1985).

Environment	Water Depth (m)	PP (g C/(m^2·a))	LSR (cm/(1000a))	ARC$_{org}$ (g/(m^2·a))
Coastal Upwelling	250	250	20	13
Coastal Upwelling	2500	250	15	2
Coastal Non-Upwelling	250	150	10	3
Coastal Non-Upwelling	2500	150	10	0.7
Open Ocean	3000	50	5	0.07
Open Ocean	5000	50	1	0.01

Table 19: Primary productivity in surface waters (PP), linear sedimentation rate (LSR) and accumulation rate for organic carbon (ARC$_{org}$) for different environments according to a compilation of Stein (1991).

Such a short duration becomes probable from the comparison of the data in Table 19 with those for the PS. With the assumptions of i) oxic bottom water conditions, ii) normal (not high) primary production rates in coastal seas and iii) a reasonable water depth of 250m, an accumulation rate of 3 (g org. C/(m^2·a)) is to be expected from Table 19, some four to six times higher than established for the PS in Southern Germany (Table 17). As anoxic rather than oxic bottom waters prevailed during most of the time of PS deposition, even higher accumulation rates for the same water depth and the same palaeoproductivity rates are reasonable and the discrepancy becomes even greater. It can only be resolved, if the duration of PS deposition was much shorter, i.e., about 100,000 - 250,000 years rather than 2.5 million years. An additional argument for a shorter duration of the PS is the low sedimentation rate for the PS of about 10m in 2.5 million years or 4mm in thousand years. Although this number would be higher for a decompacted PS, it is evident that it is as low as sedimentation rates on recent sediment starved distal ocean ridges such as Broken Ridge in the central Indian Ocean (Littke et al., 1991a). For a continental shelf, much higher sedimentation rates can be expected.

A shorter time of PS deposition implies a longer time covered by erosional events. According to the sedimentologic and geochemical data on cores from northern and southern Germany, the only major erosional disconformity present in all profiles is the one at the top of the PS. Based on the assumption of correct data on the duration of the entire Toarcian (7.5 million years, Harland et al., 1990) it its therefore assumed that the duration of the erosional event that finalized PS deposition lasted much longer than the deposition itself. Based on the calculations and discussions above a duration of PS deposition of less than 250.000 years can be expected. The following erosion accordingly lasted more than seven million years. This reasoning is based on the assumption that originally not much thicker PS was deposited or, in other words, that not more than a few metres of PS were eroded.

4.9   Depositional history

In northern Germany, the well-laminated character and absence of current features throughout most of the PS indicate deposition below current or wave base. According to Galloway and Hobday (1983: 144-158) a maximum depth of about 200 metres has been documented for storm waves, but would likely be less in the shallow Toarcian sea. Water depths greater than wave base generally persisted during the entire PS deposition. The lack of current activity fits with the palaeogeographic reconstructions which show that the Toarcian sea was extensive, shallow and well surrounded by land (Fig. 24), where development of high amplitude wind-generated surface waves would be impaired. Perhaps the Baltic Sea (Manheim, 1961) is an existing example of this land - sea configuration. Black shale deposition was inhibited where water depth was shallow enough for waves to reach the sedimentary surface. Thus, palaeo-relief on the basin floor also controlled the thickness of the PS with elevated locations showing thinner profiles.

It is of special interest whether the coquina (fossil hash) layers present in southern but also in northern Germany represent long times of non-deposition of fine material. In this case, bottom waters would very probably be oxic, benthic activity would have been reestablished and organic matter in directly underlying black shales would be more severely degraded than in the rest of the PS. No such effect was found for the coquina separating Units I and II in northern Germany (see more extended discussion in Littke et al., 1991b); therefore it is tentatively concluded that these layers represent rapid depositional events, i.e., short time intervals. This is important for the reasoning on the duration of PS deposition.

In northern Germany, after coquina deposition, fine-grained sedimentation (Units II and III) resumed, but sediment composition became clay-dominated with higher sedimentation rates indicating an increase in supply of clay material. It is reasonable to assume that this clastic debris was transported by rivers into the Posidonia sea. The bioclastic rudite, i.e., Lias zeta facies, at the top of the PS might result from shoal-water conditions established when mud deposition (Units II and III) finally built up to water depths coinciding with current base. If the shales were prograding sequences, termination of PS deposition (i.e., establishment of shoal-water) may have occurred progressively later from margin towards the basin centre. Careful regional lithofacies and biostratigraphic studies are needed to test these ideas.

In southern Germany, initial black shale deposition which is represented by the thin "Tafelfleins" layer ended after only a short time (Fig. 25B). Overlying sediments ("Aschgraue Mergel", Fig. 23) are light-coloured, bioturbated, and relatively poor in organic matter which is hydrogen-poor. Sedimented marine organic matter was severely degraded, probably in the presence of oxygen which was provided by turbulent bottom waters, eventually storm waves. At Wickensen, no similar event occurred, i.e., there are no organic matter-poor sediments above the bottom black shales (Fig. 25A). In southern Germany highest C_{org} and HI-values in the shales just below and above the "Unterer Stein" ("Unterer Schiefer", Fig. 23) indicate the most undisturbed anaerobism for the uppermost *tenuicostatum* and lowermost *faliciferum* zone. Similarly, in Wickensen the carbonate-rich sediments at the base (lower 7m of the PS) which probably represent *tenuicostatum* and *falciferum* zones (compare Loh et al., 1986 and Littke and Rullkötter, 1987) reveal exceptionally high HI values and may be representative of consistent reducing conditions.

The trends of decreasing C_{org} and HI values above the "Unterer Stein" limestone are interpreted to result from greater fluctuations of the oxygen content at and near the sediment/water interface which is also evident from the more variable lithologic appearance, e.g., the occurrence of fossil hash layers and the "Oberer Stein" limestone. The simplest explanation for this trend is shallowing of the water perhaps due to build-up of the black shale. No similar trend of decreasing C_{org} and HI values is present within the PS in northern Germany. However, recently obtained results on $Ni^{2+}/(Ni^{2+} + VO^{2+})$-ratios in porphyrins and C_{org} isotopic composition indicate a progressive trend towards less reducing bottom water conditions for the upper part of the PS in well Wickensen (Sundararaman et al., 1991). Thus, a shallowing of the water and gradually increasing rates of organic matter degradation can be assumed for both regions (northern and southern Germany) during the late Upper Toarcian (*bifrons* zone).

The lithologic (absence of bioturbation) and geochemical homogeneity of the PS in well Wickensen compared to the four South German PS profiles supports the view that deposition took place in deeper water in northern Germany. The similarity of average maceral composition of all PS profiles is regarded as an indication that the same biota provided the bulk of the organic matter at the various locations. Greater water depth in northern Germany led to a more continuous black shale deposition not interrupted by C_{org}-poor, bioturbated layers. Furthermore, the higher and more uniform hydrogen index values in northern Germany indicate better organic matter preservation than in southern Germany.

If the above conclusions are true, the most C_{org}-rich facies was deposited in the central, deeper parts of the epeiric Lower Toarcian sea, e.g., in northern Germany. Such a pattern with greatest C_{org}-contents in deepest parts of sedimentary basins is also observed in recent lakes and seas with stagnant bottom water (Huc, 1988a,b). Furthermore, in this type of environment organic matter is often well preserved, i.e., C_{org}-and HI-values are high (see Chapter 3). This coincidence leads to the suggestion that stagnant conditions also existed during PS deposition (cf. Jenkyns, 1988). Necessary prerequisites for the generation and maintainance of stagnant bottom water are 1) a suitable basin configuration and topography inhibiting rapid water circulation and 2) physical differences between bottom and surface water with respect to temperature or salinity which favour oxygen depletion in bottom water (e.g., Dean and Arthur, 1989).

Hypersaline bottom water due to erosion and leaching of Permian salt domes was proposed as primary cause of salinity stratification in northern Germany during the early Toarcian (Jordan, 1974). There is some recent support for this idea based on boron concentration in clays and microscopical observations (Janicke, 1990), but organic geochemical criteria for hypersalinity such as phytane dominance over pristane (ten Haven et al., 1987) are missing. Whereas pristane/phytane ratios below 1, often below 0.7, are reported from most sediments deposited under hypersaline water (ten Haven et al., 1987), average ratios at Wickensen vary between 1.4 and 1.7 for different lithologic units within the PS (Table 14) and average ratios are in the same range in southern Germany (Rotzal, 1990). Some evidence for hypersaline conditions is provided by the occurrence of gammacerane which was detected in the PS in northern Germany (Rullkötter and Marzi, 1988), but not in southern Germany (Bauer, 1991). According to recent studies, gammacerane may, however, be regarded as an indicator of mild hypersaline conditions (de Leeuw and Sinninghe Damsté, 1990) only at high concentration levels (Mello et al., 1988). As there is no organic geochemical indication for the presence of hypersaline bottom water, it is concluded that low salinity surface water was overlying bottom water of normal marine salinity. The presence of salinity stratification as the prime control of Lower Toarcian black shale formation is supported by phytoplankton association (Prauss and Riegel, 1989) indicating brackish surface water. Marine water could flow into the epeiric sea from the south (Tethys) and possibly also from the north; the inflowing freshwater may be derived mainly from the northeast (Scandinavia; Fig. 24), where higher rainfall than in the south is assumed (Parrish et al., 1982). Accordingly, the greater accumulation rates for all major sedimentary components of the PS in northern Germany are interpreted as the effect of high clay and nutrient input and higher bioproductivity in surface water in the vicinity of the Scandinavian landmass. The higher C_{org} contents and HI values in northern Germany are tentatively explained by better preservation of organic matter due to the existence of a more stable water stratification. This more stable stratification may be due to greater water depth as proposed above or - alternatively - caused by greater salinity contrast between surface and bottom water than in southern Germany.

Rates of planktonic carbonate and marine organic matter (phytoplankton) production and supply of detrial clay are the principal controls of the primary composition of the PS. The change from marlstone to shale facies (Units II and III in Wickensen) is best explained by an increase in clay supply. Carbonate - clay compositional differences between shales and marlstones and the greater laminae thickness of the shale facies both indicate that terrigenous influx may have increased by as much as a factor of two. The positive correlation between clay and kerogen implies that phytoplankton productivity in the Lower Toarcian sea was linked to terrestrially derived nutrients. In contrast, carbonate constituents act as a dilutant to the organic matter content. Hence, it appears that organic tissue ('soft-parts') of carbonate plankton contributed little to the total organic matter content of the PS.

As indicated by lack of burrowing, abundance and preservation of hydrogen-rich organic matter, and pronounced sulphate-reducing bacterial activity, severely anoxic conditions prevailed below the sediment - water interface throughout PS deposition. It is equally clear that a well oxygenated photic zone supporting planktonic (and nektonic) organisms persisted during deposition. It is more difficult to assess the conditions of bottom waters. Carbon - sulphur relationships indicate that euxinic conditions, i.e., H_2S-rich waters, did not extend far above the depositional interface. The abundance of benthic bivalves suggests that there were periods when bottom waters were tolerable to epibenthos, i.e., dysaerobic (Table 2). Similarly, C/S relationships indicate that Unit II bottom waters were sufficiently aerated to oxidize and recycle H_2S released from bottom sediments. It is concluded that although bottom waters were generally anoxic, the anoxic zone was relatively thin, so that on occasion aerated waters could extend to the sediment - water interface in response to small perturbations in the controlling mechanisms.

In summary, transgression and resulting development of a widespread shallow epeiric sea, possibly with a low-salinity cap and a stagnant (current-free), more saline and generally anoxic bottom zone, are major controls of PS deposition. They are, however, not sufficient to account for the high accumulation of organic matter within the sequence. As recently advocated by Pedersen and Calvert (1990), primary (i.e., algal) productivity, perhaps in generous amounts, is also an essential prerequisite. Primary productivity rates are, however, extremely difficult to establish for the PS in view of the uncertainties about the duration of this depositional event. Comparison with literature data on primary productivity (Tables 17, 18, 19) reveals that the duration of PS deposition was much shorter than previously assumed. With the 'classic' assumption of a duration of two to three million years, very low primary productivity rates result which are not sufficient to maintain oxygen-deficient bottom waters over the large area covered by the Posidonia sea.

5.   PETROLEUM GENERATION

5.1   Petroleum source rocks

Sedimentary rocks rich in organic matter are regarded as source rocks for oil and gas. The concept of source rocks is necessary in petroleum exploration since there is clear evidence that commercial petroleum accumulations did not form in situ in reservoir rocks. Most reservoir rocks are porous sandstones, limestones and dolomites that contain only small amounts of primary organic matter, i.e., finely dispersed organic matter of biologic origin. Thus, oil and gas must be derived from other sources and have migrated into reservoir rocks which were originally almost devoid of organic matter. Consequently, petroleum source rocks are defined as sedimentary rocks containing high amounts of insoluble organic material (kerogen) that can partly be transformed into liquid and gaseous hydrocarbons during burial and heating.

Effective source rocks are generally fine-grained sedimentary rocks with high concentrations of hydrogen-rich kerogen. The classical source rock concept defines a petroleum source rock as either a siliclastic sedimentary rock containing more than 0.5% (by weight) organic carbon or a carbonate with more than 0.3% organic carbon. This definition of a lower C_{org}-limit is based on empirical observations on C_{org}-contents in oil-bearing and non-oil bearing basins by Russian scientists (Ronov, 1958; see Tissot and

Welte, 1984: 674). This lower C_{org}-limit for petroleum source rocks is related to the ratio of generated oil to the storage capacity of the rock: in sediments with low concentrations of organic matter, hydrocarbons will be generated during burial and heating just like in organic matter-rich source rocks, but petroleum expulsion will not take place, because the storage capacity of the rock is not exceeded and conditions and processes necessary to initiate expulsion were not fulfilled or did not take place (see Durand, 1987). Hence, when considering petroleum source rocks the generation of hydrocarbons and their primary migration (=expulsion out of the source rock) cannot be separated.

Organic material finely disseminated in sedimentary rocks is called "kerogen", if it is insoluble in organic solvents (Durand, 1980:25). Chemically, it is to a great extent derived from complicated macromolecules present in the lipid or lignin-rich fractions of biomass that form many resistant parts of organisms such as membranes, inner cell walls of woody material, cuticles, spores, pollen, algae, bacteria, etc. These parts of decayed organisms tend to be incorporated in sediments as particulate organic matter. Much of the kerogen is therefore visible as microscopically small particles (=macerals). Their size varies between 0.002 and 1mm (or more) and they can often be attributed to precursor groups like wood or algae. Dependent on the vastly different chemical composition and structure of precursor materials, kerogen is cracked to liquid or gaseous hydrocarbons or remains "inert" during catagenesis. In organic geochemistry, the term diagenesis is restricted to the uppermost sediments, in which almost no thermally induced petroleum generation takes place, whereas catagenesis and metagenesis describe the temperature/pressure regimes in which liquid petroleum and gas, respectively, are formed.

The bulk of the organic matter in sediments and sedimentary rocks consists of finely disseminated kerogen particles. Kerogen is the most abundant naturally occurring organic material in the earth's crust. All known fossil fuels in economic deposits, (gas, oil, tar sands and coals) are collectively several orders of magnitude lower in absolute amounts than kerogen (Fig. 32). In other words, with respect to the global distribution of organic carbon in the earth's crust, economic deposits of oil, gas and tar sands are rare anomalies rather than the rule. The occurrence of these anomalies does not only depend on the presence of good source rocks, but also on the right level of maturation and on the availability of migration pathways and traps. In other words ideal geological conditions are required for the accumulation of petroleum. This in a very broad sense is the complexity of oil and gas exploration.

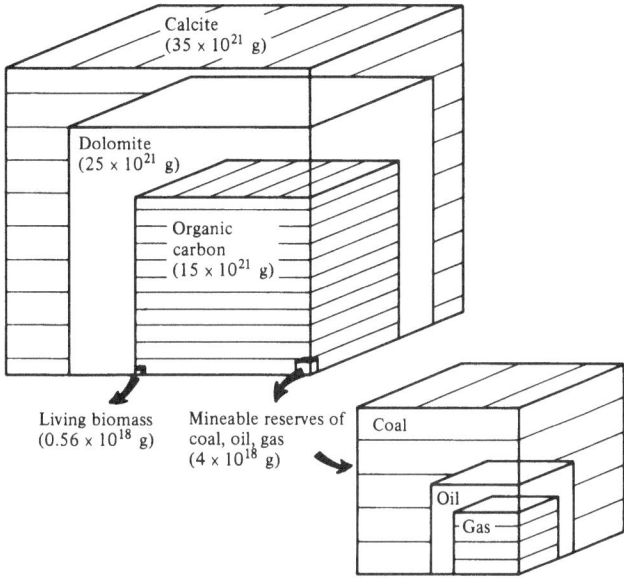

Fig. 32:   Estimated mass of organic carbon and of carbon fixed in dolomite and calcite in the earth's crust. Only a small percentage of the organic carbon is fixed in producible reserves of coal, oil, and gas or in living biomass (from Littke and Welte, 1992 using data of Berner and Lasaga; 1989).

5.2   Maturation and hydrocarbon generation

The process of organic matter transformation with increasing temperatures is called maturation. With respect to petroleum generation, organic matter may be immature (=premature), mature or overmature. The involved reactions are irreversible. In order to describe the maturation process, different maturation parameters like vitrinite reflectance, spectral fluorescence of alginite and sporinite or ratios of specific biomarker molecules have been established. They allow a definition of the state of maturity of organic matter in a specific sedimentary rock. Although these parameters are temperature- and time-sensitive, contrary to a common belief they do not describe the wide variety of processes of petroleum generation rigorously. During petroleum generation a great number of different reactions take place. Therefore, petroleum generation is not restricted to one single temperature threshold, but extends over a range of temperatures. Thus, there is a range of activation energies for petroleum generation as a bulk process which cannot accurately be described by the changes of specific maturity parameters that obey different laws of reaction kinetics. The usefulness of maturity parameters in predicting petroleum generation is further limited by the significant differences in kerogen structure and composition in different source rocks. For example, sulphur-rich kerogen will be transformed at much lower temperatures than sulphur-poor kerogen (Baskin and Peters, 1992; Rullkötter et al., 1990). Hence, rather precise knowledge on the chemical composition and structure of kerogen in a specific source rock is required to obtain reasonable information on temperature history and timing of petroleum generation from maturity data.

About 30 years ago, Jüntgen and coworkers (e.g., Jüntgen and Karweil, 1962) studied the kinetics of gas (methane) release from coals using experimental pyrolysis. During this procedure a sample is heated at a well-defined heating-rate in an inert atmosphere to high temperatures and the gas generation is recorded. In more recent years, similar analytical devices were used to study petroleum generation from immature source rocks (e.g., Espitalié et al., 1988). This experimental procedure offers the principal advantage over the use of maturation parameters that petroleum generation can be predicted or reconstructed within the framework of such important variables as type of organic matter and temperature history. At the present time, the most sophisticated kinetic models describe petroleum generation from a specific source rock by a range of activation energies and frequency factors, considering the cracking of kerogen to oil and gas and of oil to gas (e.g., Ungerer, 1990, Ungerer et al., 1988). In this context, however, it should not be forgotten that heating rates during laboratory pyrolysis experiments usually vary only between 0.1 and 25°C/min and are by about 10-12 orders of magnitude lower under natural geologic conditions. As time for petroleum generation in experiments is much shorter, temperatures are much higher, usually in the range between 300°C and 500°C instead of between 60°C and 200°C during natural petroleum generation. The reliability of any kinetic data derived from experiments clearly depends on whether natural petroleum generation obeys the same principle reactions as experimental maturation in the laboratory. At the present time, exploration concepts should include maturation assessment using both maturation parameters selected according to the specific geologic conditions in a sedimentary basin and kinetic experiments on petroleum generation from possible source rocks in the study area.

5.3   Optical maturity parameters

By means of incident light microscopy, three types of organic particles, so called maceral groups (Stach et al., 1982), are generally recognized: inertinite, vitrinite, and liptinite (syn. exinite). Inertinite is charcterized by high reflectance, liptinite by low reflectance and vitrinite by intermediate reflectance. Vitrinite and inertinite are the products of predepositional and diagenetic transformations of tissues of higher land plants, which primarily consist of cellulose and lignin. The exact conditions required for either inertinite or vitrinite precursors to be incorporated in sediments are not known, but a low degree of degradation is probably a prerequisite for vitrinite occurrence (e.g., Styan and Bustin, 1983). Liptinites are derived from hydrogen-rich plant remains such as sporopollenin, cutin, resins and waxes, or suberin

(Teichmüller, 1982; Given, 1984) or from marine or lacustrine phytoplankton such as green algae (Teichmüller and Ottenjann, 1977).

Vitrinite reflectance was first used to measure accurately and rapidly the rank of coals. The observation that vitrinite particles are ubiquiously present in most sedimentary rocks led during the last two decades to an intensified use of vitrinite reflectance both as 1) maturity parameter (predicting the stage of oil generation mainly from macerals other than vitrinite) and as 2) a calibration tool for numerical simulations of temperature histories in sedimentary basins (Waples, 1980).

Research during the last two decades revealed that organic matter in rich petroleum source rocks is mainly composed of alginite derived from phytoplankton (e.g., Hutton et al., 1980). Alginite is, however, only preserved under favourable conditions and, accordingly, is present only in relatively thin source rock intervals. This is one major reason, why changes in the optical properties of liptinite, especially of alginite, cannot generally be used quantitatively as maturity indicators in thick stratigraphic columns. Optical maturity parameters other than vitrinite reflectance may nevertheless provide important and reliable qualitative information on maturity levels.

5.3.1 Identification of macerals at different stages of dispersion

Only organic petrographic methods allow the comparison of optical properties of specific types of organic constituents at different maturation stages, whereas geochemical methods commonly measure properties of the mixture of all types of organic constituents present in a rock. The major weakness of organic petrographic maturity evaluation is, however, the subjectivity of identification of the correct maceral group or maceral used for optical measurements.

Organic petrography using incident light microscopy on polished whole rock blocks is derived from coal petrography (Teichmüller, 1986). In coals, maceral groups can easily be identified because: 1) organic particles are generally large, and 2) occurrence of clay and other minerals which have a low reflectivity - similar to liptinite - does not interfere with the proper identification of macerals. Thus, in coals the three maceral groups can easily be distinguished based on their reflectivity. Furthermore, it is possible to identifiy <u>different</u> vitrinite and inertinite macerals in coals and to restrict reflectance measurements to one vitrinite maceral only (telocollinite) rather than to measure reflectance values on all vitrinites. The various types of vitrinite are described in great detail by Stach et al. (1982) and are defined by internal structure and external shape. Their mean reflectance varies in the same depth interval (< 3m) as shown in Tab. 20 for seven coal seams (Littke, 1987). The data indicate that mean reflectances of different primary vitrinite types may differ considerably (generally up to 15% difference).

Clearly the identification of macerals and maceral groups in sedimentary rocks other than coals depends on the size of organic particles. Generally, the size of vitrinite and inertinite particles which are derived from higher land plants depends on the length of transport and the intensity of mechanical degradation. Particle sizes generally decrease with increasing distance from land areas (Fig. 7). In open marine deep sea sediments, terrigenous organic particles are usually smaller than 20 μm. With increasing degree of mechanical destruction and grain size reduction, the identification of the various types of vitrinites (Tab. 20) and inertinites becomes more and more difficult. In practical cases, the low-reflecting population of the organic particles other than liptinite is often defined as vitrinite. This choice of the low reflecting population has the general advantage that inertinites or recycled vitrinites (=redeposited vitrinites from older, more mature strata) are not mistaken as unrecycled vitrinites. On the other hand, this procedure is misleading, if high-reflecting liptinites (e.g., bituminite) or solid bitumen particles are measured; in this case the low-reflecting population does not belong to the vitrinite group and reflectance values are lower than those of "true" vitrinite. Experienced petrographers therefore use - in addition to the reflectance - the morphology (external shape) and surface structure of the particles to identify vitrinites.

The identification of liptinite macerals was much enhanced by incident light fluorescence microscopy introduced into source rock studies by Jacob (1961; see Teichmüller, 1986). This method allows the identification of liptinite macerals, such as alginite, sporinite, resinite, etc., in most sedimentary rocks. In some sediments, however, petrographic characterization of organic matter is virtually impossible, probably due to the small particle size. Examples are black shales from offshore Peru with 2 - 8 % organic carbon but almost no microscopically identifiable organic particles (ten Haven et al., 1990). In such a case it is reasonable to conclude that organic particles are smaller than the maximum microscopic resolution, i.e., smaller than about 1 μm.

It should be noted that identification of macerals is greatly complicated if pulverized rocks or kerogen concentrates - treated by various acids to remove mineral matter - are studied rather than oriented whole rock sections (see Fig. 1). The major reason is based on the fact that the form of organic particles is of great help for their classification. If the particles are microscopically studied in all kinds of obscure angles rather than in consistent orientation (usually perpendicular to bedding), classification becomes more difficult. In general, kerogen concentration is only helpful for microscopic studies if the mineralogical composition of the sediments does not allow the preparation of well-polished whole rock sections or if organic particles are extremely rare.

Vitrinite-"type"	Midg. 2	Hagen	P4Un.	N	C2	A	Zoll.3
Depth (m)	790	890	1240	1267	1402	1425	1471
Vitrinite (total)	0.65	0.71	0.88	0.88	0.95	0.99	0.92
Telocollinite	0.66	0.72	0.89	0.90	0.95	0.99	0.92
Desmocollinite	0.60	0.67	0.83	0.81	0.92	0.96	n.d.
Thin fibres	0.63	0.70	0.86	0.83	0.92	0.99	n.d.
Corpocollinite	0.68	0.73	0.94	1.00	n.d.	1.01	n.d.
Vitr. betw. cut.	0.61	0.66	0.81	0.84	0.91	0.93	0.85
In claystone	0.64	n.d.	n.d.	0.85	n.d.	0.96	0.93

Table 20:   Mean vitrinite reflectance for different kinds of vitrinite in seven coal seams from the Ruhr area, western Germany (Littke, 1987). Each of the seven intervals is less than 3 metres thick. Thin fibres are less than 50 μm thick. Vitrinites between cutinites, and vitrinite groundmass (desmocollinite) show lower values than telocollinite (vitrinite layers thicker than 50 μm). Dispersed vitrinites in claystones are characterized by reflectances similar to telocollinite in coal seams. n.d. = not determined.

5.3.2 Vitrinite reflectance

Upon maturation, vitrinites lose volatile products such as water, carbon dioxide, organic acids, and hydrocarbons (van Krevelen, 1961, Littke et al., 1989). These chemical changes are accompanied by changes of physical properties. Most important, there is an increase in vitrinite reflectance (R) which according to the Fresnel-Beer equation

$$R = \frac{(n-n_0)^2 + n^2k^2}{(n+n_0)^2 + n^2k^2}$$

is a function of the absorbtion coefficient (k) and the refraction (n) of the vitrinite particle and of the refraction of the overlying medium (n_0; usually oil with n_0= 1.518; see Ting, 1982 for more details). The increase of reflectance values with increasing temperature and burial depth is mainly an effect of the aromatization of organic macromolecules and of the condensation of the aromatic entities (Fig. 33; Schenk et al., 1990; see Béhar and Vandenbroucke, 1986). One of the advantages of vitrinite reflectance over almost all other maturity parameters is that it can be used over a wide range of diagenetic intervals, i.e., from the peat to the anchimetamorphic anthracite stage. This is one of the reasons why vitrinite reflectance served as the standard maturity parameter during the last two decades. Its use is restricted to the Post-Silurian time, because higher land plants as vitrinite precursors are only known from this period (Goodarzi et al., 1992). Interestingly, organic particles with reflectances typical of vitrinite are also known from older rocks, e.g., from the Cambrian Alum shale of Scandinavia. Their origin is not yet known, but the difference in external shape between "typical" Post-Silurian vitrinites and Cambrian vitrinites does not suggest a common origin (Plate 1E; Horsfield et al., 1992).

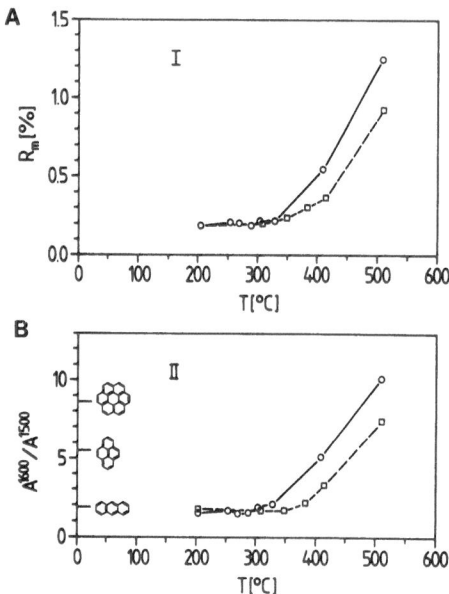

Fig. 33:   Temperature dependence of (A) mean vitrinite reflectance (R_m) and (B) of the ratio A^{1600}/A^{1500} of vitrinite aromatic stretching band areas at 1600 and 1500 cm^{-1} at heating rates of 0.1 (solid lines) and 2.0°C/min (dashed lines). The values of A^{1600}/A^{1500} for anthracene, pyrene, and coronene are indicated for comparison (from Schenk et al., 1990).

The changes of reflectance respond to temperature and the time during which specific temperatures are maintained (e.g., Karweil, 1956; Bostick, 1979; Waples, 1980; Sweeney and Burnham, 1990). Additional factors such as the chemical environment or pressure are believed to have only a minor influence on reflectance levels or to have an influence only at high stages of maturation (graphitisation; see Lyons et al., 1985). Vitrinite reflectance values are generally measured randomly, in the ordinary beam and published as mean values; this is the reason for different abbreviations (R_r, R_o, R_m) found in literature for the same type of data. At reflectance levels exceeding 1.5%, vitrinite anisotropy is increasing to an extent that mean maximum reflectances and mean minimum reflectances (R_{max}- and R_{min}-values) are determined rather than random reflectance (see Stach et al., 1982 for more details).

The validity of vitrinite reflectance as a maturity parameter predicting the stage of oil and gas generation is dependent on the similarity between the kinetics of vitrinite reflectance increase and bulk petroleum generation. Under laboratory conditions, most petroleum generation from alginite and vitrinite takes place at temperatures clearly below those at which vitrinite (and alginite) reflectance start to increase (Fig. 34; Schenk et al., 1990), whereas experience from studies in sedimentary basins indicates that petroleum generation in nature takes place at advanced reflectance levels (greater than 0.5 % Rr) corresponding to temperatures greater than 80-100°C (Quigley et al., 1987). Thus, vitrinite reflectance increase and petroleum generation seem to be dependent on heating rate in a different way. The experience that oil-generation "on average" starts at 0.5% Rr and ends at 1.3% R_r does not mean that this is true for all sedimentary basins. Especially basins which experienced high heating rates may show reflectance-generation relationships much different from the "average" (Yalcin and Welte, 1989). Therefore, geological information on burial and temperature history have to be used in combination with vitrinite reflectance data in order to predict the state of oil generation rather than a simple "oil-window" concept.

As an additional problem, it is not yet known to what extent factors other than time and temperature, such as host rock facies, influence vitrinite reflectance increase during catagenesis. In comparison with depth - reflectance trends defined by measurements on coal samples, those determined on vitrinite particles from clastic and carbonate rocks usually show the same general trend but more scatter (Fig. 35). In most cases, mean vitrinite reflectance values measured on rocks of different petrographic composition but from the same narrow depth interval do not differ by more than 15 % of the value and may be attributed to different botanical vitrinite precursors (Tab. 20) or early diagenetic processes (Buiskol Taxopeus, 1983).

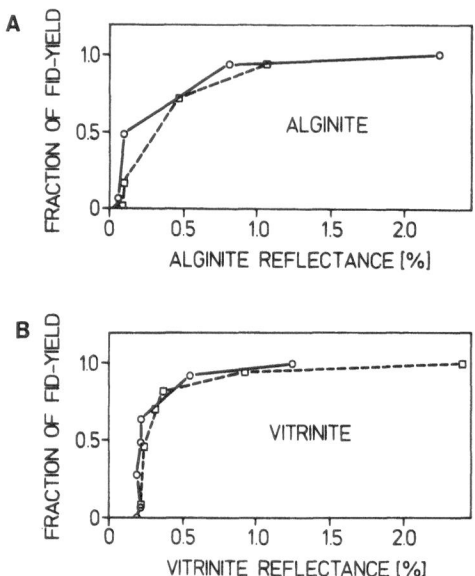

Fig. 34:   Fractions of total products generated during pyrolysis of alginite (A) and vitrinite (B) as a function of mean reflectance at heating rates of 0.1 (solid lines) and 2.0°C/min (dashed lines). The bulk of the products are generated at temperatures which do <u>not</u> cause a significant increase in reflectance (from Schenk et al., 1990).

"Suppressions" of reflectance of vitrinite in oil shales were reported by Hutton and Cook (1980) and other authors and attributed to an incorporation of bitumen from external sources, i.e., from liptinites ubiquiously present in oil shales. This explanation seems to be premature, because no significant difference in reflectance exists between data measured on extracted and nonextracted (bitumen-impregnated) specimen of the same sample. Furthermore, it was demonstrated that not even the release of the bulk of volatile products from vitrinite upon pyrolysis significantly changes reflectivity (Fig. 34), but that reflectance increase is mainly due to the rearrangement of aromatic units (Fig. 33, Schenk et al., 1990). Accordingly, differences between reflectance values of dispersed vitrinites in clastic rocks cannot be expected to depend on bitumen impregnation in a simple way. It seems possible that part of the well-known low-reflecting vitrinites in oil shales are derived from different precursors such as humic precipitates (see Chapter 3.5) than vitrinites in most other sedimentary rocks which are derived from higher land plants.

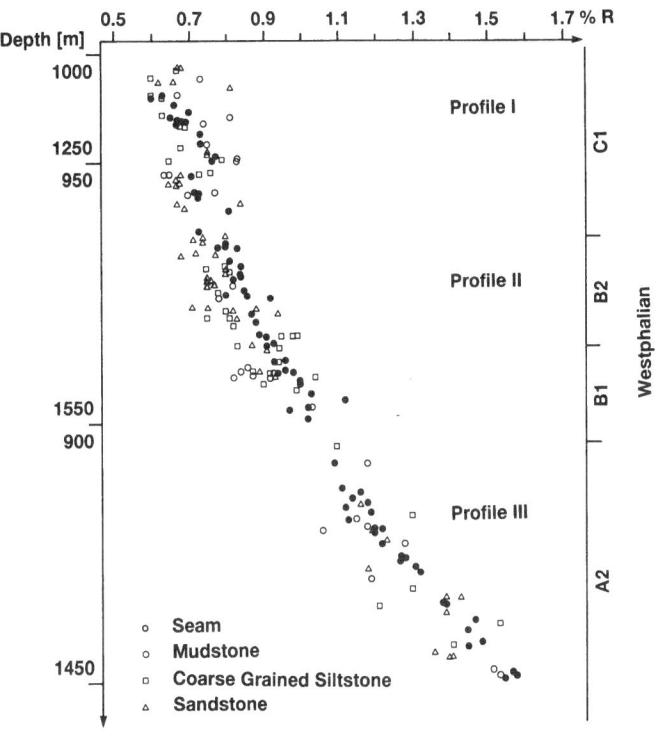

Fig. 35:   Depth and age dependence of mean vitrinite reflectance (R_r) in three wells in the Carboniferous Ruhr area, western Germany. Dispersed vitrinites in mudstones, coarse-grained siltstones, and sandstones (open symboles) show the same trend (but with more scatter) as vitrinites from coals seams (black dots; from Scheidt and Littke, 1989).

5.3.3 Reflectance of liptinite, zooclasts, and solid bitumen

Reflectance of zooclasts and organic particles other than vitrinite were used in the past to evaluate maturity, especially in pre-Devonian rocks and in other sedimentary rocks in which no vitrinite is present.

Reflectance of liptinite, e.g., sporinite, does not change significantly at low levels of maturation corresponding to vitrinite reflectance values lower than about 0.8 %. Above this level, liptinite reflectance increases more rapidly than vitrinite reflectance (Fig. 36; Littke, 1987). The major disadvantage of liptinite reflectance as a maturity parameter is the difference in optical properties between the major liptinite macerals, i.e., reflectance of resinite, sporinite, alginite, cutinite etc. differ greatly (Teichmüller, 1982). Therefore, reflectance of liptinite can serve as reliable maturity indiator only where the same type of liptinite occurs in a thick stratigraphic section or where the same type of liptinite occurs in one peculiar source rock in different areas.

Bertrand (1990) summarized data on the reflectance of zooclasts and found that "chitinozoans and graptolites show similar reflectivity and values slightly lower than vitrinite, while scolecodents show significantly lower values than that of vitrinite". Zooclast reflectance is regarded as a valuable maturity indicator in Palaeozoic rocks (Goodarzi et al., 1992).

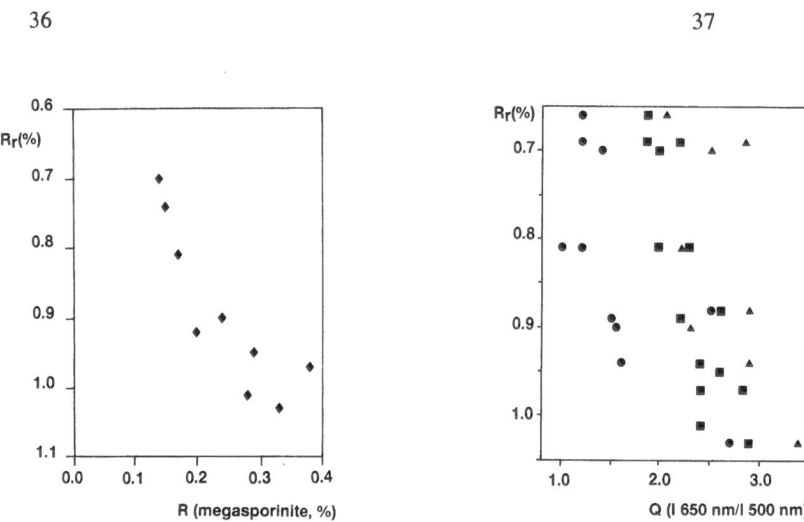

36   37

Fig. 36:   Increase of the mean reflectance of megasporinite versus mean vitrinite reflectance (R_r) in coal seams of the Carboniferous in the Ruhr area, western Germany. At R_r-levels above 0.8-0.9%, sporinite reflectance increases faster than vitrinite reflectance (after Littke, 1985).

Fig. 37:   Plot of mean red/green quotient (Q; fluorescence intensity at 650 nm divided by fluorescence intensity at 500 nm) of megasporinite (squares), cutinite (triangles), and fluorinite (circles) versus mean vitrinite reflectance (R_r) for samples from coal seams in the Carboniferous of the Ruhr area, western Germany (data from Littke, 1985).

The reflectance of solid bitumen (~ syn. pyrobitumen, migrabitumen, natural tar) has recently been advocated and discussed as an alternative to vitrinite reflectance measurements by Jacob (1989), although the great scatter of solid bitumen reflectivity at low levels of maturation (< 1% vitrinite reflectance) restricts its utility. At higher ranks, solid bitumen (see Curiale, 1986 for chemical information) is a ubiquious constituent of sedimentary rocks and its reflectance could provide important information on mature to overmature sedimentary rocks. Recent observations by Bertrand (1990) on differences between mean solid bitumen reflectance in various lithologies (sandstone, shale, limestone) and different regression lines between solid bitumen and vitrinite reflectivity for these lithologies show, however, that further studies and more data are needed before solid bitumen reflectance can be regarded as a reliable maturity parameter.

5.3.4 Coloration of palynomorphs and conodonts

The observation of progressive colour changes of liptinite macerals in transmitted light with increasing maturation led to the early development of carbonization measurements on pollen grains and spores and their application in petroleum exploration (Gutjahr, 1966; see Staplin, 1977 for historical review). Based on these observations, the "thermal alteration index" (TAI) was established as maturity parameter. One major problem of this parameter is that different liptinites and even different types of spores differ in their optical properties; accordingly, "each type of organic matter should have its own scale calibrated to hydrocarbon analysis" (Staplin, 1977). Furthermore, grain thickness greatly influences transmission, and description of color was until recently a subjective evaluation rather than a physical measurement. Nevertheless, coloration of palynomorphs was successfully used as a rapid though rough estimate of thermal maturation in many sedimentary rocks. Translucency measurements on one specific pollen genus (*Carya*; Eocene-Recent) have been related by Lerche and McKenna (1991) to thermal history via first-order time-temperature integrals. Another indicator for the maturation of organic matter is the colour of conodonts which consist of calcium fluoro-apatite. At high levels of maturation (exceeding about 1.5% vitrinite reflectance) the conodont alteration index (CAI) changes in response to temperature and time (Epstein et al., 1977). CAI and vitrinite reflectance are well correlated in various areas (e.g., Nöth, 1991; Köngishof, 1992).

5.3.5 Spectral fluorescence of liptinites

The development of a method providing spectroscopic information on individual macerals detailed enough that chemical information can be deduced, i.e., a microprobe for organic particles, is still not achieved (see Blob et al., 1988 for discussion). The most widely used spectroscopic and microscopic method to characterize organic particles in rocks is fluorescence spectroscopy (see Teichmüller, 1986 for historical review; Lin and Davis, 1988). Unfortunatelly, it reveals almost no chemical information (see Pradier et al., 1990). It is, however, an important empirical finding that most brightly fluorescing organic particles are hydrogen-rich (oil-prone) immature liptinites, and that most weakly fluorescing particles are hydrogen-lean macerals. At increasing maturation stages, fluorescence of liptinites becomes weaker and is shifted towards longer wavelengths.

From fluorescence spectra, a number of parameters were deduced as a measure of maturity, among which the most widely used are the wavelength of maximum fluorescence intensity (λmax), the spectral red/green quotient (Q), the fluorescence intensity (I), and the alteration of green fluorescence (546 nm) during irradiation. More complex and sophisticated parameters were presented by Hagemann and Hollerbach (1981), and Michelsen and Khavari Khorasani (1990). Whereas liptinite fluorescence undisputedly provides a rapid though rough estimate of maturity and can be used in almost all (except overmature) sedimentary sequences, its use as quantitative maturity parameter is severely restricted by the great difference in fluorescence properties of different types of liptinite. An example, based on measurements of various liptinites in Carboniferous coals, is shown in Fig. 37. Obviously, fluorescence colours of megasporinite, cutinite, and fluorinite change systematically with depth and vitrinite reflectance, but the fluorescence parameters of the various liptinite macerals differ considerably at a given depth interval.

5.4   Chemical maturity parameters

Most chemical maturity parameters are based on concentration ratios of two (or more) specific molecules as determined by gas chromatography (GC) or gas chromatography coupled with mass spectrometry (GC/MS). These ratios can usually be measured accurately, although several pitfalls such as coelution of different molecules exist. In contrast to optical maturity parameters, the molecular parameters have the advantage that subjective judgements by the analyst do not influence the results. On the other hand, the molecular parameters are measured on such a small fraction of the organic matter (concentrations of relevant molecules are often in the ng/g C_{org}-range) that any contamination can lead to erroneous results. Concentrations can also be affected by biodegradation (=bacterial reworking), weathering, or water washing. Furthermore, most (or all?) of the molecular maturity parameters are not only dependent on maturity but also on organic facies (=organic matter composition, kerogen type) or on lithofacies (e.g., Rullkötter and Marzi, 1988). In the following, only few examples of molecular maturity parameters are presented. A much more comprehensive overview is found elsewhere, e.g., in Moldowan et al. (1992) and, for aromatic maturity parameters, in Radke (1987, 1988). In addition to these molecular maturity parameters, bulk chemical maturity parameters such as the T_{max}-value of Rock-Eval pyrolysis (presented in Chapter 2; see Fig. 4, Tab. 16) and spectroscopic parameters (see Fig. 33) are used in organic geochemistry, but are not further discussed here (see Espitalié et al., 1977, Solomon and Miknis, 1980; Witte et al., 1988).

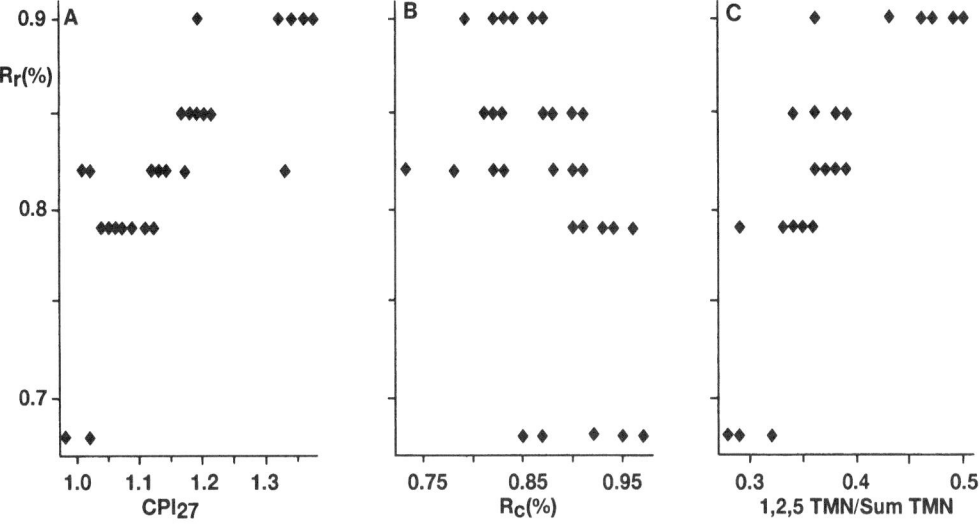

Fig. 38:   Plot of CPI (A; see text), of R_c calculated from MPI (B; see text), and of ratio of 1,2,5-trimethylnaphthalene/sum of trimethylnapthalenes (C; see text) versus mean vitrinite reflectance (R_r) for coal samples from well Nesberg 1 in the Carboniferous of the Ruhr area, western Germany (data from Littke et al., 1990).

One of the oldest and best known molecular maturity parameters is the carbon preference index (CPI; Bray and Evans, 1961; Leythaeuser and Welte, 1969) which measures the concentration ratio of n-alkanes with odd carbon numbers over those with even carbon numbers. An example is

$$CPI_{27}=n\text{-}C_{27}/(1/2(n\text{-}C_{26}+n\text{-}C_{28})).$$

CPI-values are also used for other carbon numbers in the C_{15}- C_{30}-n-alkane range. Generally, immature sediments with high amounts of terrigenous organic matter are characterized by a dominance of long-chain n-alkanes (C_{25}-C_{33}) and by CPI-values greater than one in this range (odd/even predominance). With increasing maturity, CPI-values closer to one are found. An example is shown in Fig. 38A for coals from the Ruhr area (Littke et al., 1990). In this sequence in which the organic matter is almost exclusively of higher land plant origin, the decrease of CPI-values is well correlated with the increase of vitrinite reflectance (R_r). At about 0.9% R_r, the CPI "end-value" of 1.0 is reached. At higher maturity levels, the CPI value cannot be used as maturity parameter.

In clastic sediments, in which marine or lacustrine organic matter predominates, long-chain n-alkanes are usually dominated by shorter chain n-alkanes often with pentadecane and heptadecane as the most prominent single compounds (see Tissot and Welte, 1984 for explanations). Here, a CPI-value for the C_{16}-

C_{18} range rather than CPI_{27} should be used as a maturity parameter. In many carbonate rocks and in evaporitic sequences, concentration ratios of n-alkanes are characterized by a dominance of even over odd numbered molecules (Tissot and Welte, 1984: 106-108; CPI-values smaller than 1). With increasing maturation, values close to 1.0 are found as in the case of clastic rocks. Thus CPI-values different from 1.0 indicate immature or early mature organic matter. CPI-values should be used as quantitative maturity parameter only in sedimentary sequences, in which no changes of organic facies and lithofacies occur as in the case of the coal-bearing Carboniferous sequence (Fig. 38A).

During the last ten years, a variety of maturity parameters based on aromatic hydrocarbons and aromatic sulphur compounds was established (see Radke, 1987; 1988 for review). The most widely used among these parameters is the methylphenanthrene-index (MPI, Radke and Welte, 1983) which is based on the ratio of two thermally more stable over two thermally less stable methylphenanthrene isomers. The MPI-ratio can be converted into calculated vitrinite reflectance values (R_c; e.g., Radke, 1988) as shown for coals from the Ruhr area in Fig. 38B (after Littke et al., 1990). MPI-values have the general advantage over CPI-values and most other molecular maturity parameters that they are not restricted to low maturity levels (see Radke, 1988). Another example for an aromatic maturity parameter is a ratio based on trimethylnaphthalenes (Püttmann et al., 1988) which correlates well with vitrinite reflectance in the case of the Carboniferous coal-bearing sequence in western Germany (Fig. 38C; Littke et al., 1990). This parameter and MPI-values are, however, valid as maturity parameters only in sedimentary rocks in which the bulk of the organic matter is of terrigenous origin. For sedimentary sequences with predominant aquatic organic matter, other aromatic maturity parameters have to be used (Radke, 1988).

In addition to the above described parameters which are based on quantitative GC of saturated and aromatic hydrocarbons, molecular maturity parameters based on GC/MS analyses (see Fig. 1) of more complex, cyclic compounds (steroids, hopanoids etc.) are used in petroleum exploration. While these "biomarker" parameters proved to be successful in prediction of maturation in many cases, especially in sediments which did not experience high temperatures, the reaction mechanisms became more and more disputed. For example, sterane "isomerization" and hopane "isomerization" (Mackenzie and McKenzie, 1983) are probably better described by selective degradation of the relevant sterane and hopane molecules as shown by ten Haven et al. (1992) for the latter group. Using quantitative GC/MS these authors described that the concentrations of hopane and moretane (=17β(H), 21α(H)-hopanes) are decreasing in coal-bearing strata in the relatively small vitrinite reflectance interval of 0.70-0.81% by 34% to 93%, respectively (Fig. 39). Thus, a better preservation of hopane relative to moretane rather than a reaction of moretane to hopane (isomerization) seems to be the principal cause for the maturity-related increase of the hopane/moretane ratio. While the missing knowledge of the reaction mechanism is more of fundamental than of economic interest, a more general disadvantage of most "biomarker" parameters for maturity prediction lies in the fact that they are strongly facies-dependent. For example, Rullkötter and Marzi

97

(1988) compared a variety of biomarker maturity parameters for a black shale interval and over- and underlying sediments. They found systematic differences for most parameters, i.e., values in the black shale were either smaller or greater than in the adjacent sediments, although they experienced the same temperature history.

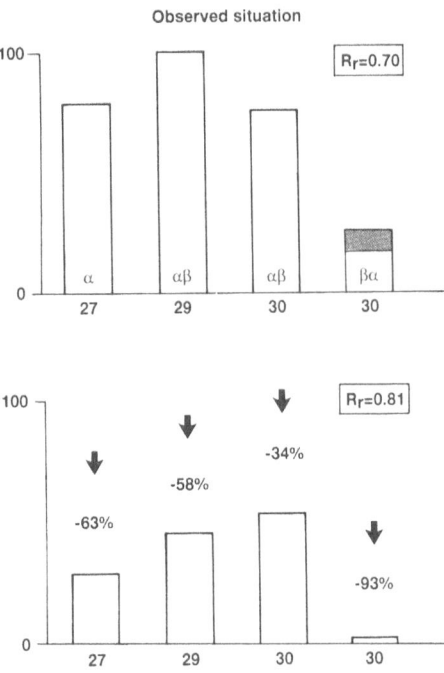

Fig. 39:   Bar graphs showing the relative concentration (highest concentration defined as 100) of selected hopanoids (bacterial-derived biomarkers) in two coal seams at maturity levels of 0.70 and 0.81 % R_r (lower graph). 27,29 and 30 are numbers of carbon atoms in the different hopanes. Numbers in the lower graph refer to "lost" percentages of the individual biomarkers. Greatest losses were established for moretane ("βα"; see text; figure after ten Haven et al., 1992).

Whereas the molecular parameters presented above are based on analyses of the C_{15+}-fraction of saturated hydrocarbons and the C_{11+}-fraction of aromatic hydrocarbons, additional maturity parameters were established for low-molecular-weight hydrocarbons with only four to eight carbon atoms (Philippi, 1975; Thompson, 1979; Schaefer et al., 1991). Examples are shown in Fig. 40 for the Posidonia Shale in Northern Germany (see Chapter 4) at different stages of maturation (Schaefer and Littke, 1988; see Schaefer, 1992 for more details). These parameters are only useful if core samples were immediately canned and sealed gas-tight directly after drilling, because of the great mobility of low-molecular-weight hydrocarbons.

Finally, it should be noted that molecular maturity parameters in contrast to optical maturity parameters are measured on the mobile fraction of the organic matter. If measured on oils, they provide a means to identify the depth and maturity range of source rocks and to reconstruct migration distances and timing (Rückheim, 1991). If impregnations by migrated petroleum are, however, mistaken as in-situ bitumen, molecular parameters will show wrong, often too high maturity levels. Currently, a combination of selected molecular and bulk geochemical parameters with optical parameters is probably the best way to establish the "right" maturity level for a rock or the "right" maturity trend for a well.

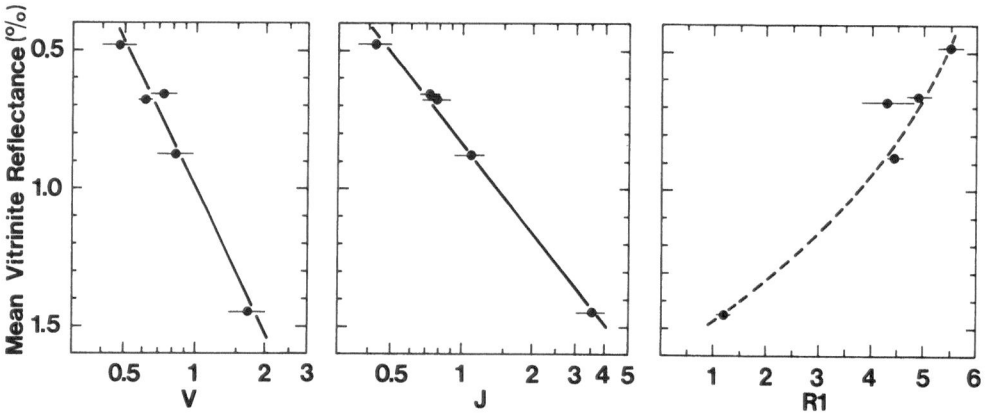

Fig. 40: Averaged maturity-sensitive low-molecular-weight hydrocarbon maturity parameters versus mean vitrinite reflectance (R_r) for five shallow cores through the Posidonia Shale, northern Germany. Bars represent standard deviations of hydrocarbon maturity parameters. V: C_7 paraffin/naphthene concentration ratio; J: (2-methylhexane + 3-methylhexane)/(1, cis-3-, + 1,trans-3, + 1,trans-2, + 1,cis-2-dimethylcyclopentane) concentration ratio; R1: (1,trans-2-dimethylcyclopentane)/(1, cis-2-dimethylcyclopentane) concentration ratio (from Schaefer and Littke, 1988).

5.5 Quantification of Petroleum Generation

The quantity of petroleum (oil + gas) that is generated in a source rock during maturation depends on its total amount of organic matter, on the thermal history, and on the chemical composition of the organic matter. In general, high H/C ratios and low O/C ratios (see Fig. 2A) characterize kerogen in excellent petroleum source rocks at low levels of maturation. Similarly, HI- and OI-values (Figs. 2B and 4) can be used to qualitatively distinguish more and less oil-prone organic matter. This classification is, however, restricted to immature rocks, because the above parameters become similar for the different types of organic matter at high levels of maturation.

The temperature-dependence of petroleum generation was illustrated earlier by "petroleum generation curves" which are plots of either burial depth or maturity parameters versus bitumen yield (e.g., Tissot and Welte, 1984). Generally, oil yields are first increasing with depth or maturity and at greater depth decreasing again, whereas gas yields are consistently increasing. A real quantification of petroleum generated in a mature source rock under natural geological conditions can rarely be achieved, because a certain quantity or fraction of the products will have been expelled. Here, particularly problematic are the more volatile products, especially methane. A reconstruction and mass balance can, however, be performed for regions where overmature to immature source rocks with a more or less identical kerogen and similar average initial amount of it are found and thoroughly analyzed.

Such a situation exists in the Hils syncline, Northern Germany, where Mesozoic sediments were influenced by a deep-seated igneous intrusion of Mid-Cretaceous age, which is known from magnetic and gravity anomalies (Reich, 1948; Hahn et al., 1976). In proximity to this anomaly ("Vlotho massif") sedimentary rocks which are older than the intrusion display features indicative of high levels of thermal alteration. These features include high vitrinite reflectance (up to R_r=4.0%, Koch and Arnemann, 1975), advanced illite crystallinities (Brauckmann, 1984), the occurrence of authigenic feldspars (Patnak and Füchtbauer, 1975), the occurrence of stoichiometric microdolomites (Richter, 1985), and high levels of conodont alteration (Nöth, 1991).

Borehole	$CaCO_3$ (%)	R_r (%)	C_{org} extr. (%)	HI, extr. (mg hc/gC_{org})	Bitumen (ppm)	Total C_8-C_{14}. Hydrocarb. (mg/g C_{org})	Total C_2-C_7 Hydrocarb.(mg/g C_{org})
Wenzen	38.5±4.0 (10)	0.48±0.02 (3)	10.6±1.9 (20)	663±43 (10)	3982±1334 (20)	8.0±1.0 (10)	6 (22)
Wickensen	36.9±9.0 (57)	0.54±0.01 (5)	10.2±2.1 (41)	656±55 (57)	6365±1412 (41)	n.d.	9 (16)
Dielmissen	34.9±4.2 (19)	0.68±0.01 (7)	9.0±1.0 (20)	574±90 (19)	12443±2555 (20)	13.7±4.8 (6)	2 (28)
Dohnsen	36.8±3.9 (37)	0.76±0.02 (4)	8.2±1.6 (27)	480±66 (37)	10546±2454 (27)	n.d.	8 (20)
Harderode	36.6±3.3 (17)	0.88±0.04 (6)	6.8±1.0 (18)	363±38 (17)	7307±689 (18)	21.6±2.7 (6)	12 (21)
Haddessen	38.9±3.9 (15)	1.45±0.02 (4)	5.5±1.8 (18)	77±2 (15)	1617±469 (18)	7.8±1.3 (6)	4 (32)

Table 21:   Geochemical data (average ± standard deviation) of the upper less calcareous part of the Posidonia Shale in six boreholes in the Hils syncline, northern Germany. R_r-values for Wickensen and Dohnsen are based on measurements on samples from the entire cored Posidonia Shale. Numbers of analysed samples are given in parentheses. Extr. = values for solvent extracted samples; n.d. = not determined. Maximum values are listed for low-molecular-weight hydrocarbons (C_2-C_7: see Schaefer, 1992). The table is based on more data than Table 1 of Rullkötter et al. (1988a).

Along the western limb of the Hils syncline the Posidonia Shale, which is the most widespread petroleum source rock of central Europe, is exposed. Adjacent to the outcrop trend and at shallow depths six cores of the compₗete Lower Toarcian Posidonia Shale (see Fig. 22) with varying amounts of the under- and overlying mudstones were drilled with the major goal to quantify the masses of petroleum generated and expelled. The core locations Wenzen, Wickensen, Dielmissen, Dohnsen, Harderode, and Haddessen were spaced at about 10km intervals (Fig. 22) along a line of progressive increase in thermal maturation towards the northwest and the inferred position of the Vlotho massif heat source. Maturation levels based on vitrinite reflectance range from 0.48% (Wenzen) to 1.45% (Haddessen).

For the mass balance of petroleum generation in the Posidonia Shale, averaged geochemical data (Table 21; Rullkötter et al., 1988a) of the homogeneous upper part (Units II and III in Chapter 4) are the base. Prerequisite for the mass balance (see Rullkötter et al., 1988a for details) is that i) the total amount of organic matter was initially (=before petroleum generation started) the same and ii) the quality (=chemical composition, see Figs. 2A and 2B) of the organic matter was initially identical. The following arguments support the view that these prerequisites are approximately fulfilled:

1. the Posidonia Shale is a widespread sedimentary rock (Fig. 24) which extends from Britain to the former Tethys as a black shale rich in hydrogen-rich organic matter (Table 11); differences in organic carbon content and in the quality of organic matter exist, however, but only over great distances (several hundred kilometres; Chapter 4);

2. the average carbonate content of the Posidonia Shale (see Fig. 25A) is almost identical in all cores in the Hils syncline (Table 21); it is expected that significant changes in the initial organic matter concentration and quality would coincide with greater variations in mineral composition; as the mineral composition (carbonate concentration, carbonate/silicate ratio) is almost identical in all profiles in the Hils area, it may be concluded that organic matter quality and concentration initially were also similar;

3. maceral composition in the less mature cores is almost identical (Littke and Rullkötter, 1987) indicating the same biological precursors of kerogen at these locations; at higher maturities (Harderode and Haddessen) many primary macerals cannot be identified (see Chapter 6.1).

Furthermore, the clay mineral composition is unexpectedly uniform in the six cores (Mann, 1987). This observation seems to contrast with the study of Brauckmann (1984) in the same area, who found distinct changes in clay mineral composition of the "Trochitenkalk" (Triassic) with decreasing distance to the Vlotho massif. Possible explanations for this difference include i) inhibited clay transformation in the Posidonia Shale due to the presence of pore-filling bitumen and ii) a potassium deficiency, i.e., a low K/clay ratio for the clay mineral-rich Posidonia Shale; a potassium deficiency is unlikely for the Trochitenkalk which contains only traces of clay minerals. In any case, the similarity in clay mineral composition provides an additional argument for an initially identical organic matter concentration and quality of the Posidonia Shale in the Hils syncline.

Based on these assumptions and the data summarized in Tab. 21, Rullkötter et al. (1988a) published a mass balance on petroleum generation (Fig. 41), which revealed that about 43% of the mass of the initial insoluble organic matter (=kerogen) were converted to oil and gas and that an additional 14% were converted to small inorganic compounds (CO_2, H_2O, H_2S). The rest remained as residue in the source rock. This residue is hydrogen-poor as indicated by the low HI-values of the organic matter of Posidonia Shale from well Haddessen. The degree of organic matter conversion established for the Posidonia Shale is probably in the range typical for many high-quality petroleum source rocks. Different conversion pathways and products are to be expected for more hydrogen-poor, oxygen-rich organic matter as typical for coals (see Chapter 7).

Interestingly, in the case of the Hils syncline, most of the kerogen conversion already occurred before the 0.7% R_r value (Dielmissen borehole) was reached. This "early generation" may be the effect of the high heating rates of the Posidonia Shale in the northern part of the Hils syncline at the time of emplacement of the "Vlotho pluton".

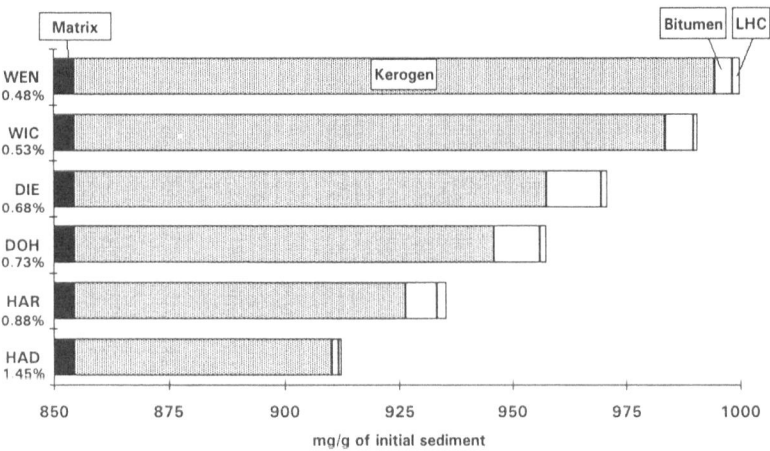

Fig. 41:   Mass balance results for petroleum generation and expulsion from the less carbonate-rich upper part of Posidonia Shale in the Hils syncline, northern Germany (see Fig. 22). All data are normalized to one gram of initial rock (Wenzen) and it is assumed that no loss or gain in the average amount of minerals occurred (extended and updated version of figure 4 in Rullkötter et al., 1988a; courtesy of Dr. B. Krooß). "Matrix" refers to the mineral matter in the Posidonia Shale, LHC are light hydrocarbons.

The great expulsion efficiency, i.e., the high ratio of expelled oil and gas over in-situ oil and gas for the mature Posidonia Shale are illustrated by a quantified "petroleum generation curve" as shown in Fig. 42 (Littke and Welte, 1992). Here, in contrast to conventional figures of this type, not only the bitumen concentration in the source rock is shown, but also the amount of expelled products, including oil, hydrocarbon gas, and inorganic gas. The in-situ bitumen yields show the typical maximum concentration at an intermediate reflectance level ("oil window"). However, the in-situ bitumen is, by far, dominated by the mass of expelled products. At Dielmissen (R_r=0.68%) 84% of the generated products are expelled. For the more mature boreholes Harderode and Haddessen, expulsion efficiencies of near 100% were calculated (Rullkötter et al., 1988a).

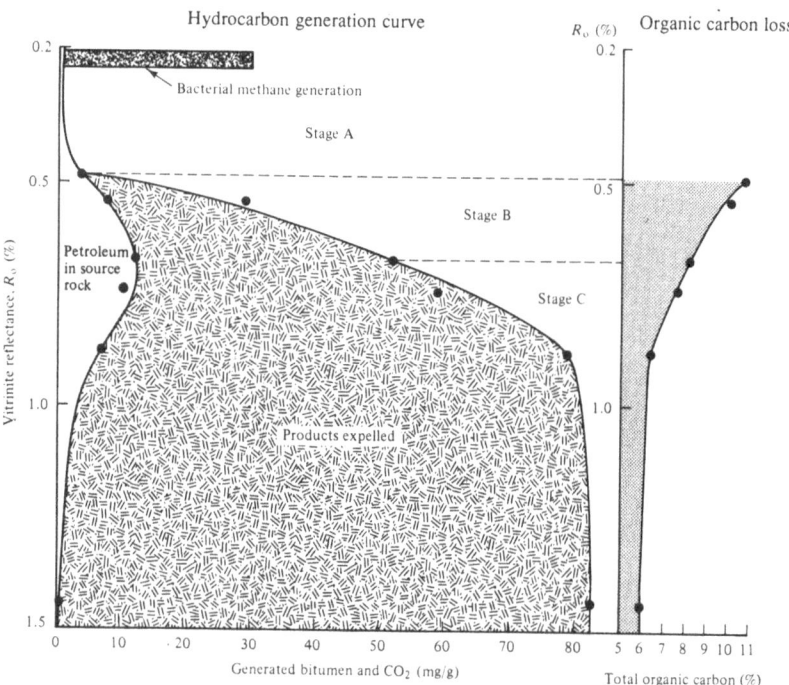

Fig. 42:   Petroleum generation and expulsion in the less calcareous upper part of Posidonia Shale in the Hils syncline, northern Germany (see Fig. 41). With increasing vitrinite reflectance (R_0) the concentration of petroleum in the source rock is first increasing and then decreasing marking an "oil window". The amount of products expelled (mainly oil and hydrocarbon gas, minor carbon dioxide) exceeds the amount of petroleum in the Posidonia Shale by far already at early stages of generation (from Littke and Welte, 1992).

The mass balance by Rullkötter et al. (1988a) is the best documented and most comprehensive published study on quantification of petroleum generation in nature, but cannot be used to get information on specific generation characteristics for other source rocks. This can be achieved by using experimental pyrolysis such as Rock-Eval pyrolysis (Chapter 2) on specific, immature source rocks and by measuring the degree of organic matter conversion in this device (see Cooles et al., 1986). Compared to the results obtained for the natural maturation of the Posidonia Shale, organic matter conversion by pyrolysis does, however, produce about 20% more volatile products and less inert residual kerogen than the natural process.

5.6   Calibration of temperature histories

5.6.1 Overview

Numerical modelling of petroleum generation (Welte and Yükler, 1981; Ungerer et al., 1984) requires crucial input data. Input data are numerical values on the parameters which control petroleum generation in sedimentary basins, i.e., numerical simulations should not attempt to handle all available information on sedimentary basin evolution, but to integrate the most relevant data. Three data sets are necessary for numerical modelling of petroleum generation:

1)   Consistent data on the burial history of a sedimentary basin based on the thickness of specific
       stratigraphic units, the duration of gaps in the sedimentary sequence and the compaction of the
       specific lithologies. Further, the entire time of basin evolution has to be divided into time steps of
       variable lengths (events) which either represent "deposition", "erosion", or "hiatus". This data set
       on the burial history is combined with the temperature evolution at the sediment/water interface
       and with the heat flow evolution as deduced from geotectonic information (see McKenzie, 1978). A
       simple example of such a data set is shown in Table 22.

2)   Calibration data to test, whether the temperature history calculated from the first data set is in
       conflict with the inherent temperature and maturity information of the rocks. For this adjustment,
       vitrinite reflectance data (see Fig. 28) and present-day borehole temperatures are the common tools,
       but a number of additional data such as molecular maturity parameters or inorganic temperature
       parameters can also be used.

3)   Kinetic data on the amounts of oil and gas generated from organic matter in response to changes of
       temperature. This kinetic data set should be specific for the organic matter in a given source rock
       within the sedimentary basin for which the temperature history is modelled, because great
       differences are known to exist between different types of organic matter with respect to the kinetics
       of petroleum generation (Tissot et al., 1987).

Early models (Lopatin, 1971; Waples, 1980) calculated temperature histories based on burial curves which neglected compaction and were based on "geothermal gradients" which were assumed to be constant over the entire time of basin evolution. In contrast, modern models (e.g., Welte and Yalcin, 1988) are founded on deterministic forward modelling techniques starting with the deposition of an oldest layer which is subsequently compacted in response to the deposition of each successive, younger layer. Furthermore, physical properties such as thermal conductivity are now assigned to each of the layers at each level of compaction. In combination with heat flow values, this information is used for the calculation of the temperature history of the different strata rather than using arbitrary "palaeogeothermal gradients" which neglect the differences in thermal conductivity of different lithologies (e.g., sandstone-salt-coal; see data in Wygrala, 1989) and variations in basin heat flow.

Event No.	Thickn. (m)	Name	Timing (ma)	Lithology	Poros. (%)	W.-depth (m)	Temp. SWI (°C)	Heatflow (HFU)
48	100	Quaternary	0-2	Shale & Sand	43	0	8	1.60
47	0	Hiatus 01	2-22	Shale & Sand	38	0	12	1.60
46	-50	Erode 45	22-23	Shale & Sand	35	0	18	1.60
45	50	Oligocene	23-37	Shale & Sand	35	30	18	1.60
44	-144	Erode 43	37-42	Shale, calc.	31	0	20	1.70
43	200	Eocene	42-47	Shale, calc.	25	30	20	1.80
42	0	Hiatus 02	47-58	Limest, marly	34	10	20	1.70
41	-9999	Erode 40	58-70	Limestone	37	0	24	1.60
40	150	Maastr.-Campanian	70-83	Shale & Limest.	37	30	24	1.60
39	-61	Erode 27	83-84	Shale calc.	35	0	22	1.60
38	-9999	Erode 28	84-85	Shale calc.	35	0	22	1.60
37	-9999	Erode 29	85-86	Shale calc.	35	0	22	1.60
36	-9999	Erode 30	86-87	Shale calc.	35	0	22	1.75
35	-9999	Erode 31-34	87-88	Limestone	35	0	22	1.85
34	65	Coniac.-Turon.	88-89	Limestone	35	70	22	2.00
33	65	Turonian	89-90	Limestone	35	70	22	1.85
32	150	Turon.-Cenom.	90-93	Limestone	35	70	22	1.72
31	100	Cenomanian	93-97	Limestone	35	70	22	1.65

Table 22: Example of an input of the basic geological data for modelling of temperature and burial history of well F (Chapter 5.6.2) with the PDI software of IES, Jülich. Erosional events are characterized by negative values in the thickness column (e.g., Event 46). A value of -9999 marks the total erosion of a sedimentary layer deposited during an earlier event. Porosities and temperatures at the sediment water interface (SWI) are calculated by the program based on lithology and compaction and on palaeolatitude and water depth at the sediment/water interface, respectively.

The most widely used calibration parameter for simulated temperature histories is vitrinite reflectance. For calibration, vitrinite reflectance-depth trends resulting from measurements are compared with vitrinite reflectance-depth trends calculated according to the simulated temperature histories. If large differences between calculated and "measured" vitrinite reflectance-depth trends exist, input parameters are changed in a geologically reasonable way. With these new data, a new temperature history and a new vitrinite reflectance-depth trend are calculated. The latter trend is again compared to the reflectance-depth trend which results from measurements.

One great pitfall in this procedure is the existence of different methods to calculate reflectance. Four of these methods will be used and compared in the following sections.

- The method of Barker and Pawlewicz (1986) is based on the assumption that the maximal temperatures experienced by a sedimentary layer are sufficient to calculate vitrinite reflectance. This method is in conflict with experience from pyrolysis experiments (e.g., Schenk et al., 1990) which suggest that heating rate (or time) do influence the increase of vitrinite reflectance. According to Barker and Pawlewicz, however, the time for equilibration of vitrinite reflectance to the current temperature regime is always sufficient under geologic conditions and the duration of heating or of maintenance of maximum temperatures can therefore be neglected. In contrast to this, Quigley et al. (1987) argued in favour of a time-dependence of reflectance increase for geologic systems.

- The TTI-method after Waples (1980) and based on Lopatin (1971) calculates a "time temperature index of maturity (TTI):

$$TTI = \sum_{n_{min}}^{n_{max}} (\Delta t_n)(r^n)$$

In this equation, Dt is the time, which a sediment layer spent in a 10°C temperature interval. The index n is defined as 0 for the temperature interval between 100 and 110°C (Table 23). TTI depends on time in a linear and on temperature in an exponential way. The factor r is commonly selected to be 2 (Lopatin, 1971), because each 10°C rise in temperature is thought to double the rate of maturity increase, independent of the respective temperature interval. TTI-values are converted into calculated vitrinite reflectance values using an empirical correlation (Waples, 1980). Variations of the original TTI-method include r-factors different from 2 and the selection of basin-specific correlations between TTI and vitrinite reflectance (e.g., Horvath et al., 1988). Today, the TTI-method is overall the most widely used method to calculate maturity (vitrinite reflectance), although it does not reflect maturity well enough in the case of complex temperature histories (Quigley et al., 1987) and although it can from a chemical point of view, "not adequately model complex reactions over a wide range of temperatures and heating rates" (Sweeney and Burnham, 1990).

- The phenol kinetic method (Larter, 1989) is based on the concentration of phenols (alcohols of benzene and methyl-benzenes) released from vitrinite at different maturities during pyrolysis. These data are used to calculate the kinetics of the reaction from phenolprecursors in vitrinites (educts) to phenols (products). The reaction is described by the first order Arrhenius equation

$$K = Ae^{(-E/RT)}$$

where K is the reaction rate, A is the frequency factor, E is the activation energy, R is the universal gas constant and T is the temperature in Kelvins. Larter (1989) used a Gaussian distribution of activation energies rather than a discrete activation energy to take the heterogeneity of vitrinite into account. The phenol kinetic method is only defined for the vitrinite reflectance range between 0.45 and 1.6%, because only in this interval quantifiable amounts of phenols are released from vitrinite.

Temperature interval (°C)	Index Value n
30-40	-7
40-50	-6
50-60	-5
60-70	-4
70-80	-3
80-90	-2
90-100	-1
100-110	0
110-120	1
120-130	2
130-140	3
140-150	4
150-160	5

Table 23:   Definition of index values n after Waples (1980). See text for usage of these values.

- The Easy % R_o method (Sweeney and Burnham, 1990) is based on the rate of release of the quantitatively most important volatile products from vitrinite (water, carbon dioxide, methane, higher hydrocarbons). The release of each of these compounds is described by the Arrhenius equation (see above). The authors used a constant frequency factor A (10^{13} sec^{-1}) and a distribution of activation energies E ranging from 34 to 72 kcal/mole. Again, this wide range of activation energies should take into account "the heterogeneity of natural materials by assuming that a complex reaction is better represented, for simplicity, by a set of parallel reactions with the same A but with different Es" (Sweeney and Burnham, 1990). Based on the distribution of activation energies, a factor F (=fraction of reactant completed) is calculated which can vary between 0 and 0.85 (theoretically 1) and which is used to calculate the mean random vitrinite reflectance as $R_r=e^{(-1.6+3.7F)}$. This kinetic method is defined up to vitrinite reflectance values of 4.5%.

Thus, maturity is with few exceptions (e.g., Barker and Pawlewicz, 1986) regarded as an integral of temperature over time. This poses a problem, if maturity parameters are used to calibrate temperature histories in basin modelling, because, even if the right calculation technique is selected, different pathways exist to calculate maturity values in agreement with measured values. The number of possible solutions is greatly diminished, however, if - in addition to maturity parameters - information on maximum temperatures reached during burial is available or on the time during which a specific temperature was maintained. This type of data can be deduced from the application of geothermometres such as fission track analyses (e.g., Naeser et al., 1989) and from homogenization temperatures of fluid inclusions (next section).

5.6.2   Simulation of the thermal history of well F

Well F, situated in the western part of the Lower Saxony Basin, was selected to demonstrate, how both organic maturity parameters and fluid inclusion homogenization data can be used to calibrate numerically simulated temperature histories (Leischner et al., 1993; Leischner, in press). The first step in this procedure was the reconstruction of burial history curves for different sedimentary units as shown in Fig. 43.

With regard to the oldest Carboniferous strata (lowermost line in Fig. 43), seven important, distinct phases of burial or erosion can be defined. During phase 1 (Carboniferous) rapid sedimentation and subsidence led to a burial depth of more than 4000m by the end of the Carboniferous. This was followed by a time of erosion during most of the Permian (phase 2), by a long time of slow deposition from the late Permian to the middle Jurassic (phase 3), by another short erosion event between Dogger and Malmian (phase 4), and by a time of slow deposition from the Late Jurassic to the middle Cretaceous (phase 5; about 90 million years before present), when the oldest strata reached their maximum burial depth of more than 6000 metres. At that time, the inversion event (phase 6) occurred which is typical for the Lower Saxony Basin (Betz et al., 1987) and which led to the erosion of almost one kilometre of previously deposited sediments. The late Cretaceous and Tertiary history (phase 7) is composed of several slow deposition and erosion events which did not significantly change the burial depth of the sediments.

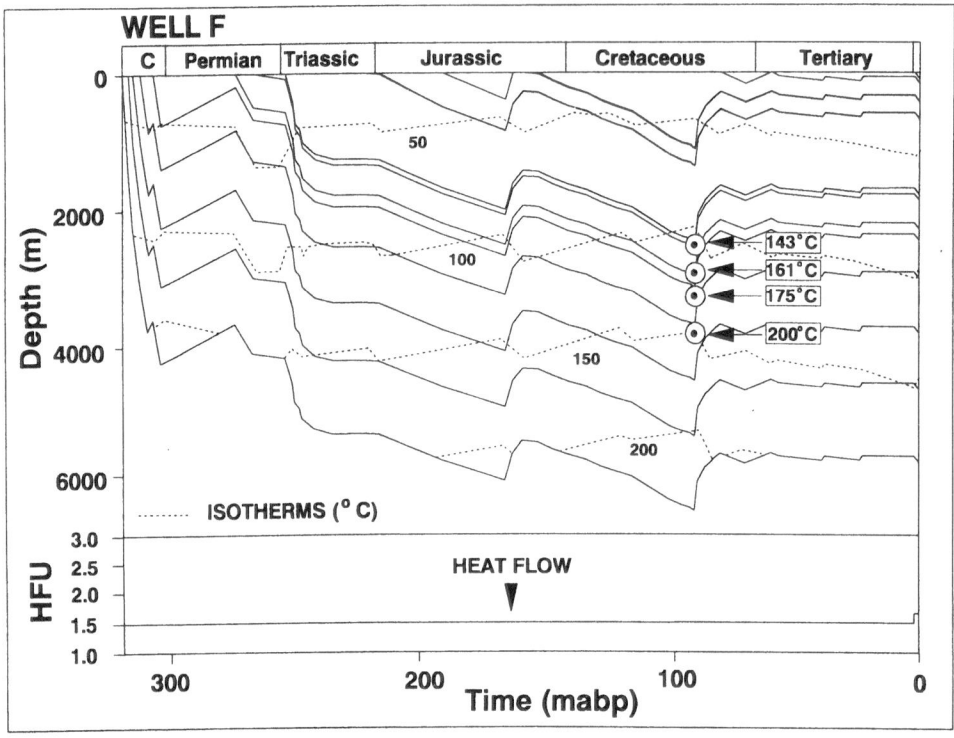

Fig. 43:   Burial history and first simulated temperature history of well F based on a constant heat flow of 1.5 HFU ($\approx$ 60mW/m^2; lower part of the figure). The temperatures indicated by circles and arrows are derived from fluid inclusions and interpreted as maximum temperatures. They do not agree with the calculated isotherms.

In a second step, a first temperature simulation was performed using 1) the burial history, 2) information on the lithology and the inherent thermal conductivity, 3) reconstructions of sediment/water interface temperatures (Table 22) and 4) an assumed constant heat flow of 1.5 HFU ($\approx$ 60mW/m^2; lower part of Fig. 43) which is typical for cratonic basins. The dashed lines in Fig. 43 show the resultant calculated isotherms for 50°C, 100°C, 150°C and 200°C.

This first temperature model was then calibrated using vitrinite reflectance data measured for this study and taken from Teichmüller et al. (1984). Results are shown in Fig. 44 together with reflectance data (R_c) calculated from methylphenanthrene-index values (MPI, see Chapter 5.4). The three sets of measured maturity data are in good agreement with each other for the entire depth range. Vitrinite reflectance is increasing from 0.55% at approximately 500m depth to nearly 2.0% at 3300m depth. T_{max} values of Rock-Eval pyrolysis (see Chapter 2) increase within the same depth interval from 428°C to 525°C. These values, although not directly convertible into vitrinite reflectance values, are in principal agreement with the reflectance - depth trend thus confirming the validity of the other maturity parameters.

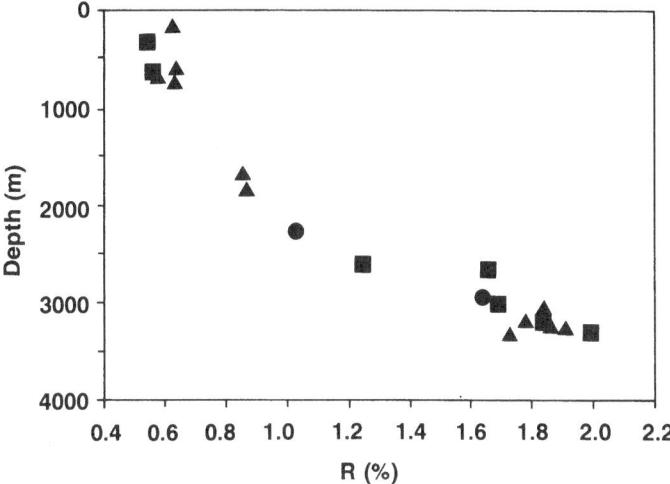

Fig. 44:   Measured vitrinite reflectance (R) data (from Leischner in press: squares, and from Teichmüller et al., 1984: circles) and vitrinite reflectance data calculated from an organic geochemical maturity parameter (methylphenanthrene index, see Radke, 1987) in well F.

Based on the first calculated temperature history (Fig. 43), a TTI- calculation of vitrinite reflectance values was performed and revealed a good agreement of measured and TTI-calculated reflectance values. In contrast, the EASY % R_0 method revealed that calculated data are consistently lower (by up to 0.4%) than measured data.

In the case of well F, in addition to maturity parameters apatite fission track data (not reported here; Leischner, in press) and data from fluid inclusion microthermometry (see Burruss, 1989) served as calibration parameters. Homogenization temperatures of fluid inclusions were measured for quartz cements in Carboniferous sandstones and for anhydrite veins in the Upper Permian (Zechstein). The data which are discussed in detail by Leischner et al. (in press) indicate maximum temperatures for different stratigraphic intervals to increase with depth from 143°C to 200°C (see Fig. 43). These maximum temperatures were probably reached at the time of maximum burial in the Mid-Cretaceous and exceed those calculated by the first temperature model (dashed lines in Fig. 43) by about 30°C (Zechstein) to 50°C (Carboniferous). As an example, the first temperature model calculates a maximum temperature of 150°C for the sandstone, for which fluid inclusion data revealed a maximum temperature of 200°C.

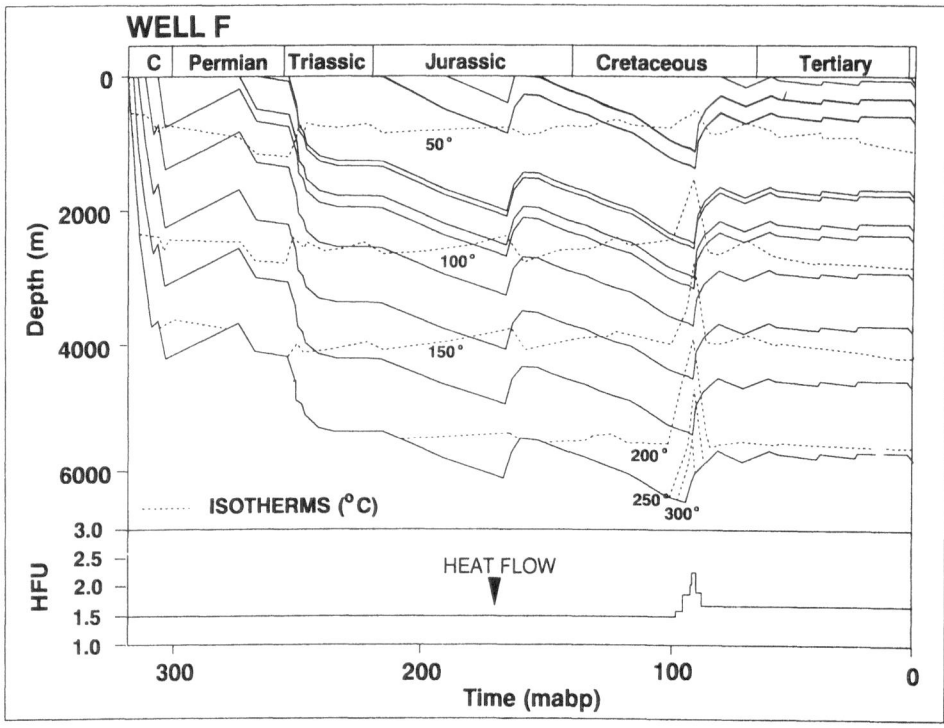

Fig. 45: Burial history and second simulated temperature history of well F based on a high heat flow at Mid-Cretaceous time (compare Fig. 43). The calculated isotherms agree with the results of fluid inclusion analyses.

Thus, a second temperature simulation was performed which is in accordance with the fluid inclusion data. This simulation was again based on 1) the burial history, 2) information on lithologies and thermal conductivities, 3) reconstructions of sediment/water interface temperatures as well as on 4) a more complex heat flow model than assumed for the first temperature model. For the second model, a constant heat flow of 1.5 HFU was selected for the time before the inversion, a heat flow maximum of 2.25 HFU during the inversion time, and a constant heat flow of 1.7 HFU for the time after the inversion (lower part of Fig. 45). The high heat flow value for the Mid-Cretaceous is believed to be related to the intrusion of a magmatic body in the vicinity of well F. Several other Mid-Cretaceous plutons ("Vlotho massif", "Bramsche massif" etc.) are believed to occur in the subsurface of the southern part of the Lower Saxony Basin and cause geophysical and diagenetic anomalies (see Chapter 5.5). The maximum heat flow value of 2.25 HFU is not unrealistically high if compared to a heat flow value of 3.5 HFU calculated by Buntebarth (1985) for the centre of the "Bramsche massif". Hence, the second heat flow model takes available geological information on the Mid-Cretaceous situation better into account than the first model.

The isotherms derived from the second thermal model are visualized as dashed lines in Fig. 45. They reveal much higher maximum temperatures than the first temperature simulation (compare Figs. 43 and 45) and are in accordance with the fluid inclusion data. For example, a calculated maximum (Mid-Cretaceous) temperature of almost 200°C results for the sandstone, for which fluid inclusion data revealed a maximum temperature of 200°C. Based on the second scenario, vitrinite reflectance values were calculated using again both the TTI-method and the EASY % R_0 approach. Only the latter calculation shows a good fit with the measured values (Fig. 46), whereas the reflectances calculated by TTI are significantly greater than measured values.

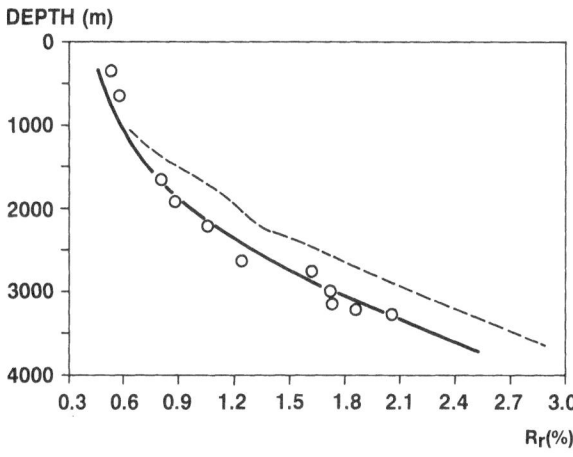

Fig. 46:   Measured vitrinite reflectance values as a function of depth and reflectance-depth trends calculated by the TTI-method (dashed line) and by the Easy % R_0 method (solid line) for well F according to the temperature history shown in Fig. 45.

In summary, it was possible to adjust the input data for well F in such a way that calculated temperature and vitrinite reflectance data are in accordance with palaeotemperature data derived from fluid inclusions and with vitrinite reflectance data, if the EASY-% R_0 calculation is selected for calculation of vitrinite reflectance values. The calibrated temperature history was used to model oil and gas generation from different sedimentary units based on kinetic data on petroleum generation (see results in Leischner, in press). The study of well F is seen as an excellent example for the need of good calibration data for realistic simulations of temperature histories and of petroleum generation.

5.6.3 Comparison of different vitrinite reflectance calculation methods

Vitrinite reflectance is the most widely used calibration parameter in basin modelling, although no standardized calculation technique exists (Chapter 5.6.1). In the following, the four methods described above are compared using burial and temperature histories of three wells in the Styrian basin of the Eastern Alps (Sachsenhofer and Littke, in press). These wells are well suited for this comparison, because many vitrinite reflectance data (Sachsenhofer, 1990, 1991) are available which are in accordance with organic geochemical maturity parameters. Another advantage is the great variability in maximum palaeo-heat flow within the Styrian basin. At some places (wells) extremely high heat flows were caused by volcanic activity during the Miocene. The three selected wells are shown in Fig. 47.

-       The Übersbach 1 well is situated at some distance from the next volcano.
-       The Pichla 1 well is situated close to a Miocene volcano.
-       The Mitterlabill 1 well is situated close to the centre of a volcano (2 km distance) and penetrated 550m of andesite. Only in the Mitterlabill 1 well exists an important disconformity, i.e., it is not known whether the time between 15 and 14 mybp (million years before present) represents a hiatus or sedimentation followed by erosion. The calculated vitrinite reflectance values should give some insight into this problem.

The Styrian basin (Fig. 47) is the westernmost part of the Pannonian basin system. The basin subsidence started 18 mybp (Ottnangian) and continued to the late Miocene. Only during the last 5-6 million years slow erosion occurred. The sediments of the Styrian basin are described in detail by Ebner and Sachsenhofer (1991).

The regional temperature history of the Styrian basin is characterized by the volcanic activity between 17 and 15 mybp (Karpatian and early Badenian). During that time shield volcanoes produced up to 1200m thick andesites. This volcanic activity influenced the surrounding sediments significantly (e.g., by an increase of vitrinite reflectance values), whereas Pliocene basaltic volcanism (see Fig. 47) had only local influence on the sediments.

The burial history for the Übersbach 1 well is shown in Fig. 48 (see Sachsenhofer and Littke for details). Vitrinite reflectance gradually increases with depth and reaches maximum values of 0.95% R_0 at the final depth of 2600m. To reconstruct temperature histories, the Easy % R_0 method was used and heat flow values were adjusted to obtain an optimal fit between calculated and measured values (see upper part of Fig. 48). Subsequently, the identical temperature history was used to calculate vitrinite reflectance with the TTI-method, the phenol kinetic method, and the Barker and Pawlewicz method (lower part of Fig. 48; see Chapter 5.6.1). The Easy % R_0 method was used as a standard, because of the good results obtained by its application in the case of well F (Chapter 5.6.2).

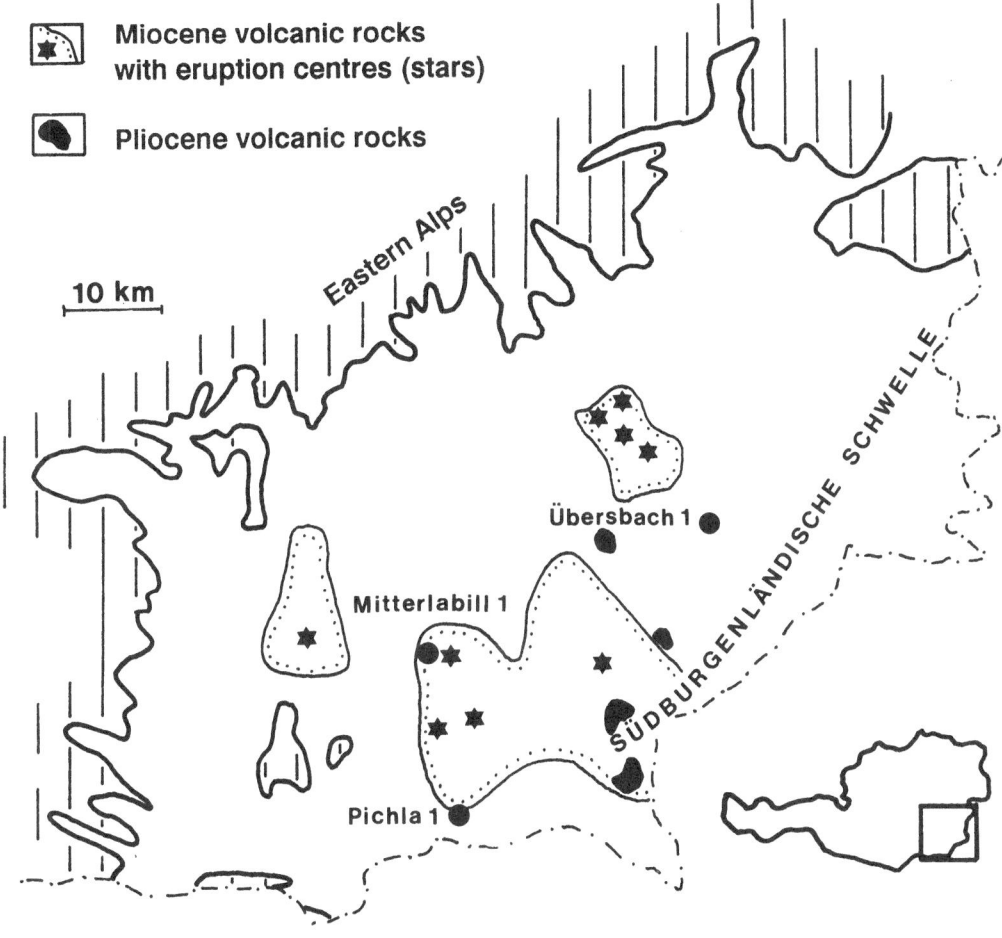

Fig. 47: Simplified map of the Styrian basin with location of volcanic rocks (after Flügel, 1988) and of studied wells (from Sachsenhofer and Littke, in press).

For well Übersbach 1, heat flows of 100 to 120 mW/m^2 are assumed for the Ottnangian to Badenian time (17 to 12 mybp) and of 80mW/m^2 for the post-Badenian time. The assumption of higher heat flow values for the early history of the Styrian basin takes a regional increase in heat flow during and after volcanic activity into account.

The results of the other three methods to calculate vitrinite reflectance do not differ much from the Easy % R_o calculation in the case of well Übersbach 1 (Fig. 48). TTI calculation is in good agreement with the measured data except for the lowermost sediments. In contrast, the Barker and Pawlewicz method reveals a good fit between measured and calculated data for the lower half of well Übersbach 1, but not for the upper half. The phenol kinetic method is in good agreement with the Easy % R_0 model within the limit of its definition (0.45 - 1.6% R_0).

In the case of the Pichla 1 well, a rapid sedimentation during Ottnangian and Karpatian was followed by slow sedimentation, hiatus, and finally erosion (Fig. 49). Vitrinite reflectance increases from 0.25% close to the surface to 2.4% near final depth revealing an exceptionally high maturity gradient.

To obtain a fit between measured vitrinite reflectance data and Easy % R_0 calculations, high heat flow values of 230 mW/m^2 were assumed for a short period (less than one million years) during the time of Miocene volcanism. For most of the later geologic history, the recent heat flow value of 80 mW/m^2 was used (Fig. 49). The value of 230 mW/m^2 chosen for the Karpatian differs only slightly from heat flows measured in the realm of recently active volcanoes (Watanabe et al., 1977).

For the selected temperature history of the Pichla 1 well, TTI is underestimating vitrinite reflectance in the lower half of the well (above 0.7 % R_0) significantly. The phenol kinetics method is in good agreement with the Easy % R_0 calculation within the limits of its definition. The Barker and Pawlewicz method overestimates R_0-values in the upper part of the well and underestimates measured data in the lowermost 500 metres.

These comparisons reveal that differences between the methods to calculate vitrinite reflectance are greater for sedimentary sequences which experienced high heat flows such as observed in Pichla 1 than for sequences with "normal" heat flow histories such as found in Übersbach 1. In contrast to the results of the simulation of well F, the comparisons performed for the Pichla and Übersbach wells do, however, not allow a validation of the four calculation methods, because Easy % R_0 was *a priori* selected as the standard method. A validation was possible in the case of the Mitterlabill 1 well.

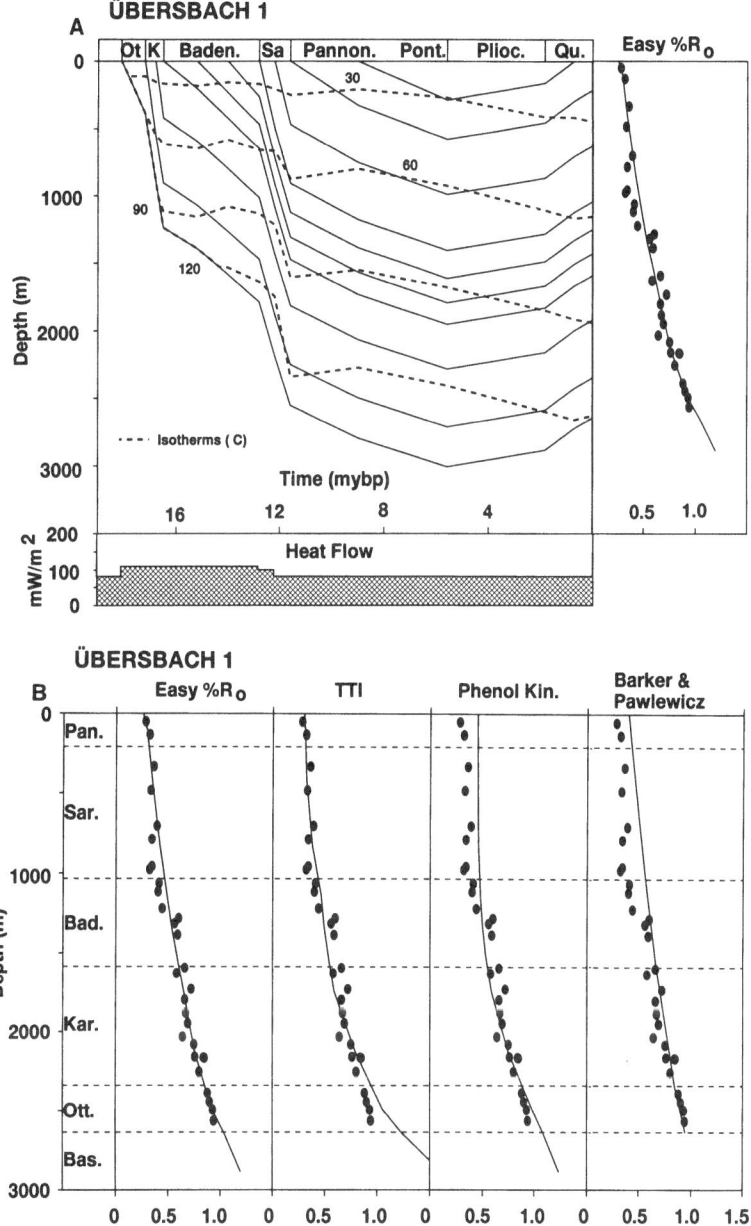

Fig. 48: (A) Burial and temperature history and heat flow model of the Übersbach 1 well. In the right part of the figure measured vitrinite reflectance values and calculated Easy %R_O-values are plotted versus depth. (B) Comparison of vitrinite reflectance values calculated by four different methods for the Übersbach 1 well according to the burial history and heat flow model in Fig. 48A (from Sachsenhofer and Littke, in press).

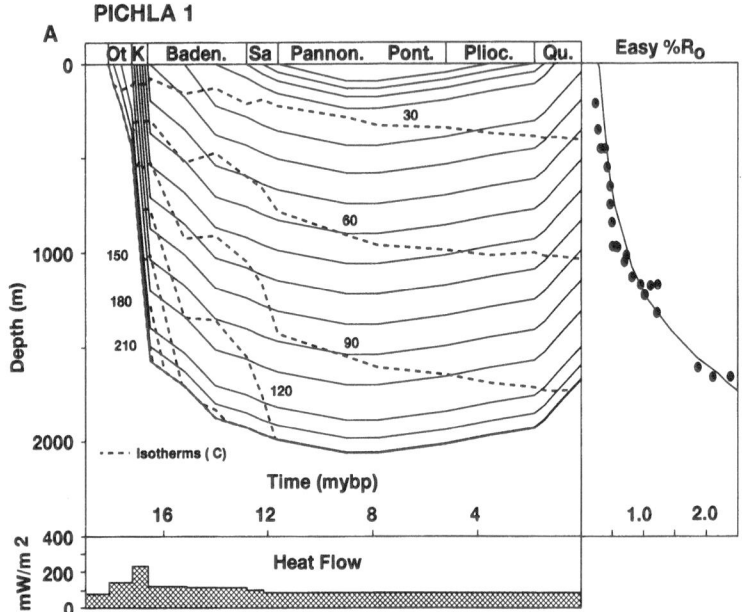

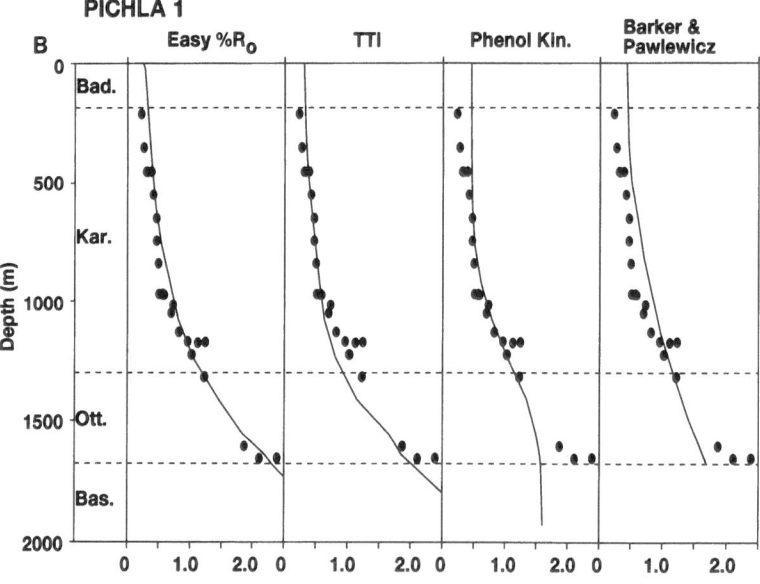

Fig. 49:   (A) Burial and temperature history and heat flow model of the Pichla 1 well. In the right part of the figure measured vitrinite reflectance values and calculated Easy %R₀-values are plotted versus depth. (B) Comparison of vitrinite reflectance values calculated by four different methods for the Pichla 1 well according to the burial history and heat flow model in Fig. 49A (from Sachsenhofer and Littke, in press).

Two possible burial histories for the Mitterlabill 1 well are shown in Fig. 50 which differ only in the interpretation of the 1 million year gap between the uppermost part of the 550m thick andesite of Karpatian to early Badenian age and the overlying late Badenian sediments. The first scenario (Fig. 50A) assumes a hiatus for the gap, whereas for the second scenario (Fig. 50B) a deposition of additional 450m of andesite and subsequent erosion was chosen. Vitrinite reflectance values are low in the sediments above the andesite and increase below from 1.0% to about 4.0% at the base of the sequence (1800m depth). For scenario 1, no satisfying fit between calculated Easy % R_0 values and measured data was achieved. The adjustment shown in Fig. 50A is based on high heat flows of >300 mW/m^2 during Karpatian and early Badenian times. With the resultant temperature history, Easy % R_0 values were calculated which are in accordance with measured R_0 data at the top and bottom of the sedimentary sequence. However, R_0-values directly below the andesite (1000 to 1500m depth) are underestimated. The selection of a higher heat flow during Karpatian and early Badenian times results in a better fit for the layers below the andesite, but in a significant disagreement between measured and calculated reflectance data near the final depth. In other word, the hiatus scenario (1) did not allow a good adjustment of Easy % R_0 data to measured data.

A good fit between the Easy % R_0 calculation and the measured reflectance data is achieved if the second burial history scenario is combined with a heat flow model which assumes maximum heat flows of 270 mW/m^2 during the Karpatian and early Badenian (Fig. 50B). If TTI instead of Easy % R_0 is used to calculate vitrinite reflectance, an equally good fit between measured and calculated data can be obtained, if the calculation is based on the erosion of at least 1500m of andesite. This latter assumption does not seem reasonable in view of the known andesite thicknesses (maximum 1200m) in the Styrian basin.

The simulation of the Mitterlabill 1 well supports the potential of the simulation techniques to gain insight into past erosional events, if the right method for the calculation of vitrinite reflectance is chosen. According to the experience with simulations of well Mitterlabill 1 in the Styrian basin and of well F (Chapter 5.6.2) in the Lower Saxony basin, the Easy % R_0 method seems to be superior to the widely used TTI method.

118

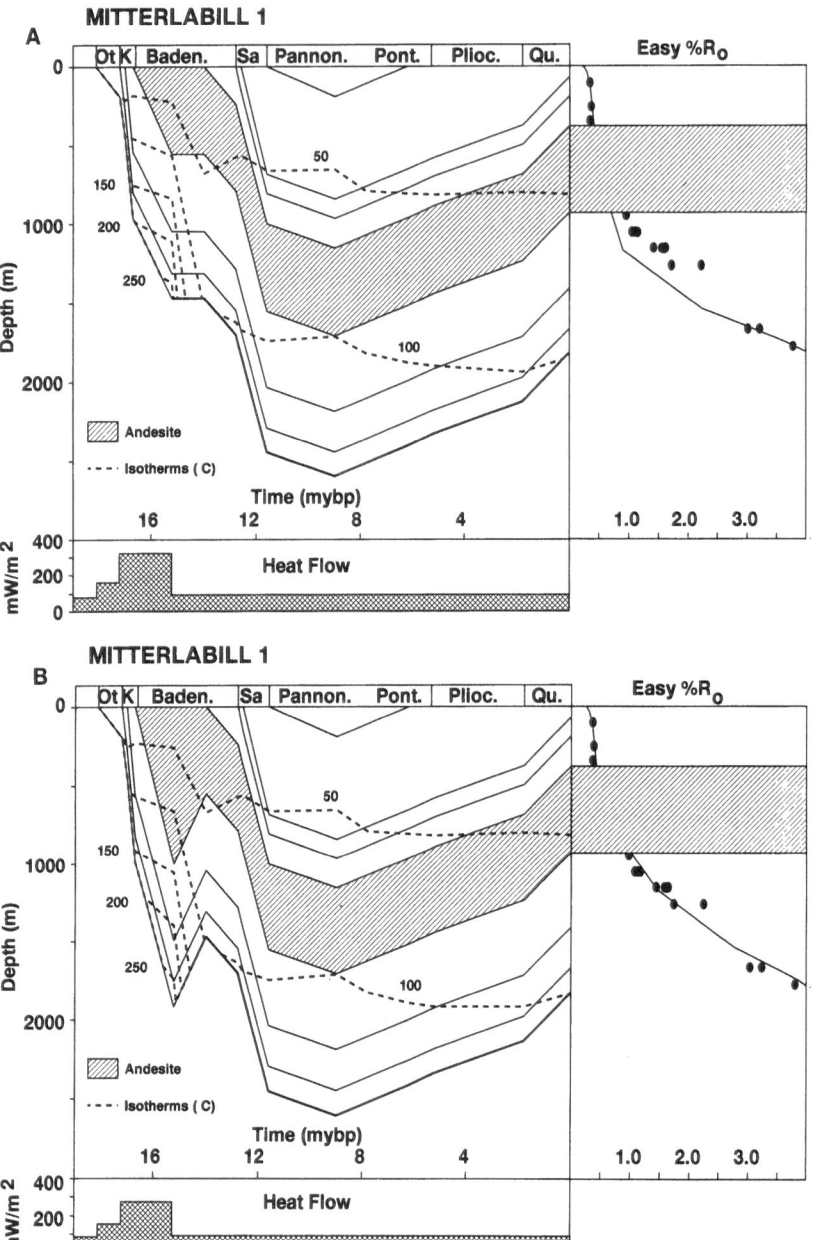

Fig. 50   Two models of the burial and temperature history and of the heat flow evolution of the Mitterlabill 1 well. The thick andesite flow overlying Karpatian sediments (in contrast to thin volcanic rocks interlayered with the Karpatian sediments) is marked by a dark signature. With the first assumption that no erosion took place during the middle Badenian (model A), the maturity trend of this well cannot be simulated. A good simulation of the maturity trend can be achieved, if an erosion of 450m during the middle Badenian (model B) is assumed. Both simulations are based on the Easy %-R_O-method (from Sachsenhofer and Littke, in press).

6. MICROSCOPIC AND SEDIMENTOLOGIC EVIDENCE FOR THE GENERATION AND MIGRATION OF PETROLEUM

The processes of petroleum generation and migration do not only influence optical and chemical maturity parameters, but also the macroscopic appearance of rocks as well as their microscopic characteristics. A few examples of this impact are presented in the following which are based on observations on source rocks and on reservoir rocks.

6.1 Evidence for petroleum generation and migration in source rocks

Effects of petroleum generation and migration in source rocks were studied in the Hils area, northern Germany (Littke and Rullkötter, 1987; Littke et al., 1988), where a total of six boreholes through the Lower Toarcian Posidonia Shale were drilled (see Chapters 4 and 5.5). These boreholes are spaced at about 10 km intervals along a line of progressive thermal maturation, ranging from 0.48% R_r to 1.45 R_r (Fig. 22). For simplicity, here only the lithologic differences of four of these six Posidonia Shale cores are reported. The cores are abbreviated as WEN (Wenzen borehole, R_r=0.48%), DIE (Dielmissen borehole, R_r=0.68%), HAR (Harderode borehole, R_r=0.88%), and HAD (Haddessen borehole, R_r=1.45%).

The general lithostratigraphic aspect of the DIE, HAR and HAD sections which have experienced progressive thermal maturation is very similar to the WEN. All show the two-fold division of a lower marlstone (54-61% average $CaCO_3$) and an upper calcareous shale facies (35-39% average $CaCO_3$), although the thickness of the shale member varies somewhat (see Table 24). Similarly, in each section the boundary between the marlstone and shale is marked by the accumulation of a thin fragmental fossil layer. Upper and lower contacts of the Posidonia Shale are as observed at the WEN locality, i.e., an erosional base with a possible paleosol developed on the underlying Pliensbachian, and an abrupt boundary with the overlying Aalenian. Similarly, on top of each of the three more mature sections, the distinctive but thin Upper Toarcian bioclastic rudite layer occurs. These lithostratigraphic similarities indicate that the original lithologic and petrographic features of the Posidonia Shale were also similar for all profiles and that any differences can be attributed to the different temperature histories (see Chapter 5.5).

The parallel laminations so prominent in the WEN core (Plate 1, I) are present but less evident at DIE, rarely discernible at HAR (Plate 1, J), and completely absent at HAD. Clearly, thermal maturation has resulted in the progressive obliteration of these primary sedimentary features. It may be of some general significance that delicate primary features, such as laminations, may not be preserved in sedimentary rocks which have experienced maturation levels of 0.9% R_r and above. The obliteration of lamination is paralleled by a darkening of the Posidonia Shale which is medium gray at WEN, dark gray at DIE and black at HAR and HAD. This is due to the impregnation of the rock by generated petroleum.

Location	Member	Thickness (m)	CaCO$_3$ (%)	R$_r$ (%)	Lamination	Fractures
WEN	Shale	11	37	0.48	+	-
	Marlstone	5	54		+	-
DIE	Shale	22	35	0.68	+	+
	Marlstone	6	55		+	+
HAR	Shale	31	37	0.88	-	+ +
	Marlstone	7	59		-	+ +
HAD	Shale	27	37	1.45	-	+ +
	Marlstone	4	61		-	+ +

Table 24: Lithologic and petrographic features of carbonate-rich shales and marlstones at the localites Wenzen, Dielmissen, Harderode, and Haddessen in the Hils area, northern Germany (after Littke et al., 1988).

Despite increasing maturation levels, mineralogic compositions of the Posidonia sections are very similar (Mann, 1987). Clay minerals, calcite, pyrite, and kerogen are the principal constituents; detrital quartz and feldspar are minor. Except for the high content of calcite in the marlstone and of clay minerals and pyrite in the calcareous shale facies, there are no appreciable mineralogic compositional variations. Clay minerals comprise up to 40% of the bulk rock. Illite is the chief component but significant amounts of kaolinite, mixed layer (illite/smectite), chlorite and also smectite are present (Mann, op. cit.). Unexpectedly, there is no indication of progressive conversion of smectite and mixed-layer clay to illite with increasing maturation. The transformation may have been inhibited by insufficient potassium either within the Posidonia Shale itself or from external, e.g., fluid sources. Another possible explanation is that the saturation of the pore volume with oil or gas was an obstacle for clay mineral transformation at the time when maximum temperatures were reached and petroleum generation occurred (Mid-Cretaceous).

Organic constituents (=macerals) in the immature Posidonia Shale were classified in four groups (see Chapter 4 for details): alginite B + liptodetrinite, alginite A, bituminite, and terrigenous macerals (vitrinite + inertinite + sporinite). At DIE, where already half of the petroleum generation capacity has been realized (Chapter 5.5, Rullkötter et al., 1988a), the amount, distribution and size of macerals are similar to those at WEN. However, the fluorescence colour of alginites is clearly different from that at WEN (Fig. 51). Interestingly, fluorescence parameters such as wavelength of maximum fluorescence intensity (λ_{max}) or red/green quotients (Q) scatter much more at DIE than at all other more and less mature locations, if unextracted rock samples are used for fluorescence measurements (Fig. 51: 560-640 nm). This scatter is much diminished, if extracted rocks are utilized for these analyses. In this context the great scatter observed in unextracted rocks can be explained as an effect of an interaction of mobile bitumen and insoluble kerogen. Different extents of bitumen impregnation of the individual alginites could cause the variability of fluorescence spectra at DIE.

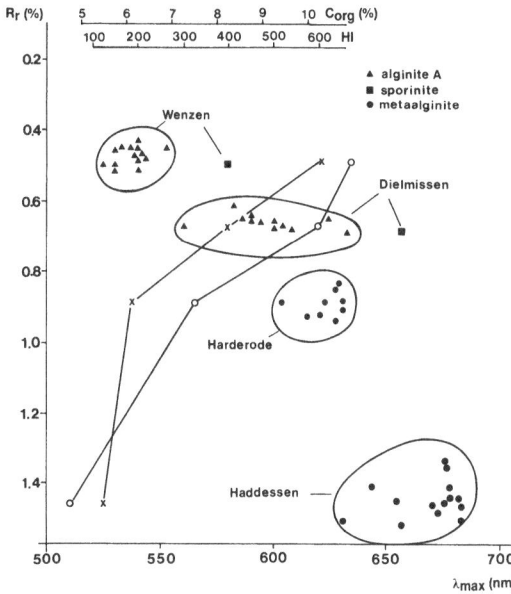

Fig. 51:   Mean vitrinite reflectance values (R_r) plotted versus mean wavelenghts of maximum fluorescence intensity (λ_{max}) of alginite A and "metaalginite" of unextracted samples from Posidonia Shale in the Hils area. The scatter of fluorescence parameters at DIE becomes much smaller, if extracted rock samples are used for spectral fluorescence measurements (after Littke and Rullkötter, 1987). Crosses mark average total organic carbon (TOC; C_{org}) values and circles mark average HI-values.

At HAR and HAD (mean R_r-values of 0.88 and 1.45%, respectively) few fluorescing alginites and no bituminites are present (Fig. 52), but micrinite and secondary, discrete, unstructured liptinites occur. These secondary liptinites are here designated as metaalginite. They are of a similar size as primary alginites B. In contrast to primary alginites, they are of irregular shape and show no preferred bedding-plane orientation. As nearly the entire hydrocarbon generation potential has been utilised and about 45% of the probable original organic carbon content is lost at HAR (Figs. 41 and 42; Rullkötter et al., 1988a), it can be calculated that primary alginites have lost about half of their original mass by oil and gas generation. Alginites seem to be able to lose about 25% of their original mass (DIE-situation) without severe morphological changes, whereas further progress of hydrocarbon generation reactions seems to result in an abrupt collapse of alginite to metaalginite (HAR-situation). Only silicified Tasmanales-alginites remain morphologically preserved, but lose their fluorescence at higher maturation levels.

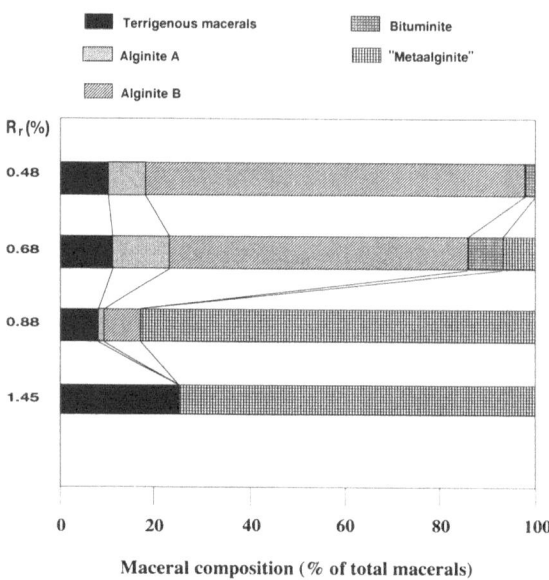

Fig. 52:   Change of maceral composition as an effect of maturation in Posidonia Shale from the Hils area (data from Littke et al., 1988).

The Cambrian Alum Shale in Scandinavia can also serve as an example for significant morphological transformations of macerals as an effect of petroleum generation. Although samples from this rock were not taken in the same systematic way as in the case of the Posidonia Shale, they provide another interesting maturity series described in detail by Horsfield et al. (1992). In the case of the immature Alum Shale, organic particles are a mixture of dominating alginite roughly similar to the alginite B of the Posidonia Shale and of non-fluorescing particles with typical spindel-shape (Plate 1, E; see also photos in Bharati and Larter, 1991) and with vitrinitic reflectance. At high levels of maturation, the primary alginite macerals are not present, but a network of finely dispersed, brightly reflecting organic matter (solid bitumen, pyrobitumen, see Jacob, 1989) is the principal organic component (Fig. 53, Plate 1, F and G). It should be noted that this complete transformation of alginite into petroleum and a new solid constituent can also be observed, if alginite is treated by laboratory pyrolysis. The residue of this experiment consists of large, highly reflecting, blocky pieces of organic matter (Plate 1, H) which have no morphological or optical similarity to the original organic matter from the Alum Shale.

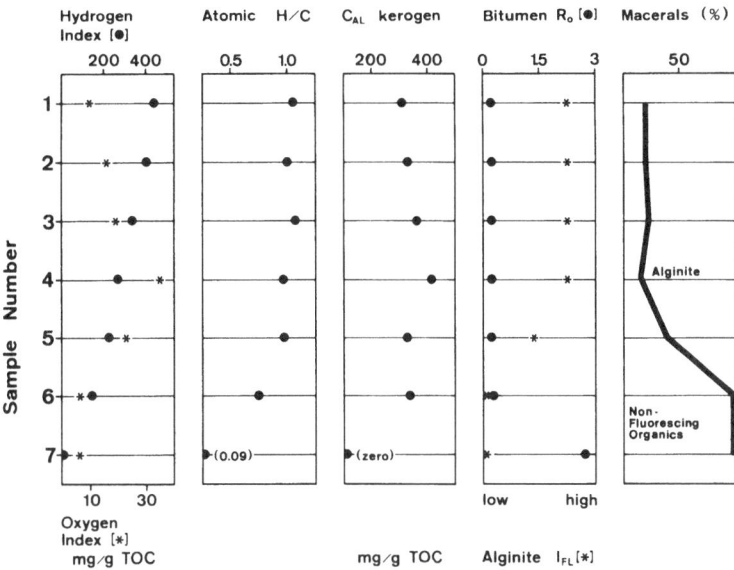

Fig. 53:   Geochemical and petrologic data on seven samples from the Cambrian Alum Shale, Scandinavia. With increasing maturation (= increase in sample numbers), HI-values, OI-values, atomic H/C ratios, aliphatic carbon in kerogen as determined by infrared spectroscopy (mg/g rock), reflectivity of "vitrinitic" material and solid bitumen (R), and fluorescence intensity of alginite as well as the ratio of fluorescing over non-fluorescing macerals change more or less systematically. The non-fluorescing organic matter at high maturation stages (samples 6 and 7) is a catagenetically formed kerogen network (= solid bitumen; after Horsfield et al., 1992; see Bharati and Larter, 1991).

Besides colour changes and maceral transformations, fracturing also accompanies petroleum generation and migration in source rocks. In the case of the Posidonia Shale, only a few vertical fractures are present in the WEN core. In striking contrast, each of the more mature cores (Table 24) displays distinctive macro- and microfractures (Plate 2, A-C) which are generally oriented parallel to the bedding-plane and which are up to 2cm thick. As an exception, a small limestone within the marlstone-unit contains a macrofracture at an acute angle ($\approx 70°$) to the bedding-plane thus indicating that the orientation of cracks is influenced by material properties (Gutjahr, 1983) of the fractured medium. Macrofractures (>200μm) are generally filled with fibrous calcite crystals which are oriented perpendicular to the fracture-boundaries and which often contain brightly fluorescing fluid inclusions trapped between the crystal margins (Plate 2, B and C). The boundaries of many of the macrofractures are commonly digitate with inclusions of wall-rock shale which are often in subparallel orientation. Microfractures, less than 200μm thick, are empty or filled with solid, non-fluorescing bitumen (Plate 2, A) and/or calcite. Consistently, the bitumen occupies the borders, whereas calcite is within the central parts of the fractures. Often, large macrofractures occur in those parts of the source rock which contain many microfractures; where microfractures are rare or absent, no macrofractures occur (see Brady, 1974). Horizontal fractures are often interconnected by thin, irregular, more or less vertical fractures.

The restriction of the fractures to those sections (DIE, HAR, and HAD) which have experienced advanced maturation (R_r=0.68% and above) is viewed as evidence that the fractures are directly related to the thermal history of the Posidonia Shale in the Hils area. Furthermore, no bedding-parallel fractures occur in the underlying Pliensbachian or overlying Aalenian mudstones, despite the fact that these rocks have experienced the same thermal history. However, in contrast to the very high initial organic matter content of the Posidonia Shale (av. ≈10.7% C_{org} at Wenzen), the adjacent shales contain little organic-matter of hydrogen-poor composition (average near 1% C_{org}, Rullkötter et al., 1988a). This difference suggests that the exclusive occurrence of these fractures in the Posidonia Shale is not only related to the temperature history but also to the generation of oil and gas which the Posidonia Shale experienced during progressive levels of maturation (see Leythaeuser et al., 1988a for more discussion).

More evidence in support of a relation of maturation and thermal stress on one hand and fracturing on the other was provided by oxygen isotope data and homogenization temperatures of fluid inclusions (Jochum et al., 1991; Jochum, in press). The oil-bearing fluid inclusions in the Posidonia Shale at Harderode predominantly contain a liquid hydrocarbon phase and hydrocarbon gas. Less commonly, one-phase (liquid hydrocarbon) inclusions were observed in the fracture fillings. Aqueous inclusions are absent. Using UV-excitation the fluid inclusions show bright fluorescence colours: yellow, brownish and light-blue. The different fluorescence colours may indicate liquid hydrocarbons of different maturity levels. McLimans and Videtich (1987) postulate a gas/condensate composition for blue fluorescent fluid inclusions. The hydrocarbon gas bubbles of two-phase fluid inclusions do not fluoresce. Using the classification from Roedder (1984) two different types of two-phase oil-bearing fluid inclusions can be differentiated in calcitic fractures at Harderode:

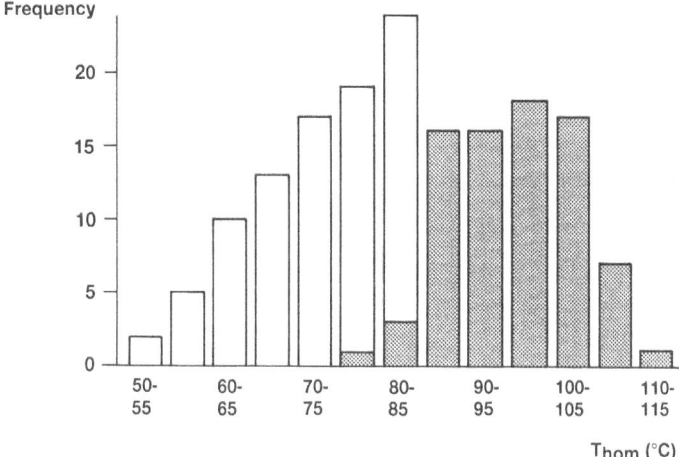

Fig. 54:   Histogram of oil-bearing fluid inclusion homogenization temperatures in calcite cement of horizontal (parallel to bedding) fractures in Posidonia Shale at Harderode, northern Germany (after Jochum et al., 1991). Primary inclusions are represented by the darker shaded parts of the histogram.

1)   A primary type, which was entrapped in the calcite crystal during initial filling of the bedding-parallel fractures. These inclusions are <5-20µm in diameter and homogenize in the temperature range of 75-115°C (Fig. 50).

2)   Secondary inclusions which are more frequent than primary inclusions and are entrapped in sealed microcracks within the calcite-filled fractures. Late tectonic movements along the fracture walls (shear stress phenomena) caused these microcracks. Secondary inclusions are <5-10µm in diameter and homogenize in the temperature range of 50-85°C (Fig. 50).

The homogenization temperatures of primary and secondary inclusions are interpreted as minimum trapping temperatures. A pressure correction of homogenization temperature data is necessary, because the Toarcian strata of Harderode were probably buried to about 1700m (Düppenbecker et al., 1991) at the time of fracture filling. Using P-V-T-data from Narr and Burruss (1984) for hydrocarbon phase systems, pressure-corrected temperatures of 145-165°C could be estimated for primary fluid inclusions. Agreement is good with numerical simulation studies carried out by PDI software which indicate a maximum temperature of 140°C for the Mid-Cretaceous intrusion time, when petroleum generation took place (Düppenbecker, 1992).

An increasingly heavier oxygen isotope composition (with much scatter; data in Jochum, in press) of the calcite in the fractures towards higher maturation levels, i.e., from WEN to HAD, also indicates that the calcite-filled fractures were opened in response to high temperatures, although no absolute temperatures could be calculated from these values. Nevertheless, the oxygen isotope data are regarded as an important argument supporting that recrystallization and precipitation of calcite in horizontal fractures took place at the time of intrusion emplacement and petroleum generation (Mid-Cretaceous), because no great differences of temperature are assumed for other times in the Posidonia Shale history at the four sites. Carbon isotope data on calcite from horizontal fractures and Posidonia Shale at different levels of maturation suggest that the source for the carbon in the fracture-calcite was the primary carbonate of the Posidonia Shale (Jochum, in press).

According to Secor (1965), horizontal open fractures can occur only when pore-fluid pressures exceed the vertical load (see Rouchet, 1981). Reconstructions indicate relatively shallow burial depths (less than 2 kilometres) for the four sections during their thermal alteration and generation process (Leythaeuser et al., 1988a). Hence, it is conceivable that pore pressure exceeded lithologic load and that the prominent bedding-plane calcite-filled fractures developed in response to overpressuring caused by oil and gas generation or by other processes related to the temperature event in the Mid-Cretaceous (see Jochum, in press). The fractures may have acted as important expulsion conduits as indicated by the presence of solid bitumen (see photos in Littke et al., 1988) and as argued by Leythaeuser et al. (1988a) on the basis of geochemical data of fractured and unfractured samples. It should be noted that the intense fracturing described above was observed in a very impermeable source rock which was rapidly heated. This may have resulted in an impedance of fluid escape (Tissot and Welte, 1984: 302) and an abnormal overpressuring. Future investigations will show to what extent the described kerogen transformation, fracturing, and hydrocarbon migration phenomena are similar to those which took place in sedimentary basins with lower heat flows and higher permeabilities in source rock intervals.

If compared to fracture orientation in other mature source rocks which are less clay-rich and less laminated (e.g., the Brown Limestone of Egypt), the horizontal fracture orientation in the Posidonia Shale is probably a special effect of the textural peculiarities of the rock. In the Brown Limestone, which is a brown-black, massive, fine-grained carbonate source rock, fractures almost perpendicular to bedding were observed (Plate 2, F) which are filled by calcite with abundant, fluorescing fluid inclusions.

Fractures related to petroleum generation or migration are not expected to occur in source rocks which are coarser grained and contain permeable conduits for petroleum migration. An example is the Kimmeridge Clay from the Brae field in the British North Sea (see Leythaeuser et al., 1988b for geochemical data). At this location, the Kimmeridge Clay consists of interlayered alginite-rich, fine grained clay and of quartz-rich silt layers on a millimetre or centimetre scale (Plate 2, D and E). No fractures were observed in samples from this rock, although it already reached the petroleum generation stage. This can easily be explained by the presence of the silt layers which can serve as pathways for pore pressure release and petroleum migration.

Plate 1.   A) "Bright particles" with dark inclusions in a Pennsylvanian black shale (Little Osage Shale) of the U.S. midcontinental area. White reflecting particles are pyrite grains and framboids. Polished surface, reflected white light, 1cm = 30 μm; B) Same view as A, but in a fluorescence mode (uv-light excitation); C) Lacustrine Messel Shale (Eocene, Germany consisting of yellow algal layers and red detritic layers. Thin section, 1 cm = 0.7 mm; D) Messel Shale consisting of bright, yellow fluorescing layers composed of small algal remains and darker detrital layers which are mixtures of siliclastic material, land plant-detritus (vitrinite and inertinite; black particles) and minor algal remains. Polished surface, fluorescence mode, 1 cm = 30 μm; E) Spindel-shaped organic particle of vitrinitic reflectance in immature Cambrian Alum Shale, Scandinavia. White reflecting particles are pyrite grains and framboids. Polished surface, reflected white light, 1 cm = 30 μm; F) Organic matter-network in mature Cambrian Alum Shale. Polished surface, reflected white light, 1 cm = 30 μm; G) Organic matter network in overmature Cambrian Alum Shale, Scandinavia. Polished surface, reflected white light, 1 cm = 30 μm; H) Large organic particle derived from pyrolysis of concentrated kerogen from the Alum Shale. Polished surface, reflected white light, 1 cm = 30 μm; I) Immature Posidonia Shale (Wenzen borehole) from the Hils area, northern Germany. Macroscopic view, 1cm = 2 cm; J) Mature Posidonia Shale (Harderode borehole) from the Hils area, northern Germany. Macroscopic view, 1 cm = 2 cm; K) Solution seam consisting of clay and organic matter in the dolomitic Permian Bone Springs formation, New Mexico, USA. Polished surface, fluorescence mode, 1 cm = 150 μm; L) Fluorescing organic particles in a solution seam in the dolomitic Permian Bone Springs formation, New Mexico, USA. Polished surface, fluorescence mode, 1 cm = 15 μm.

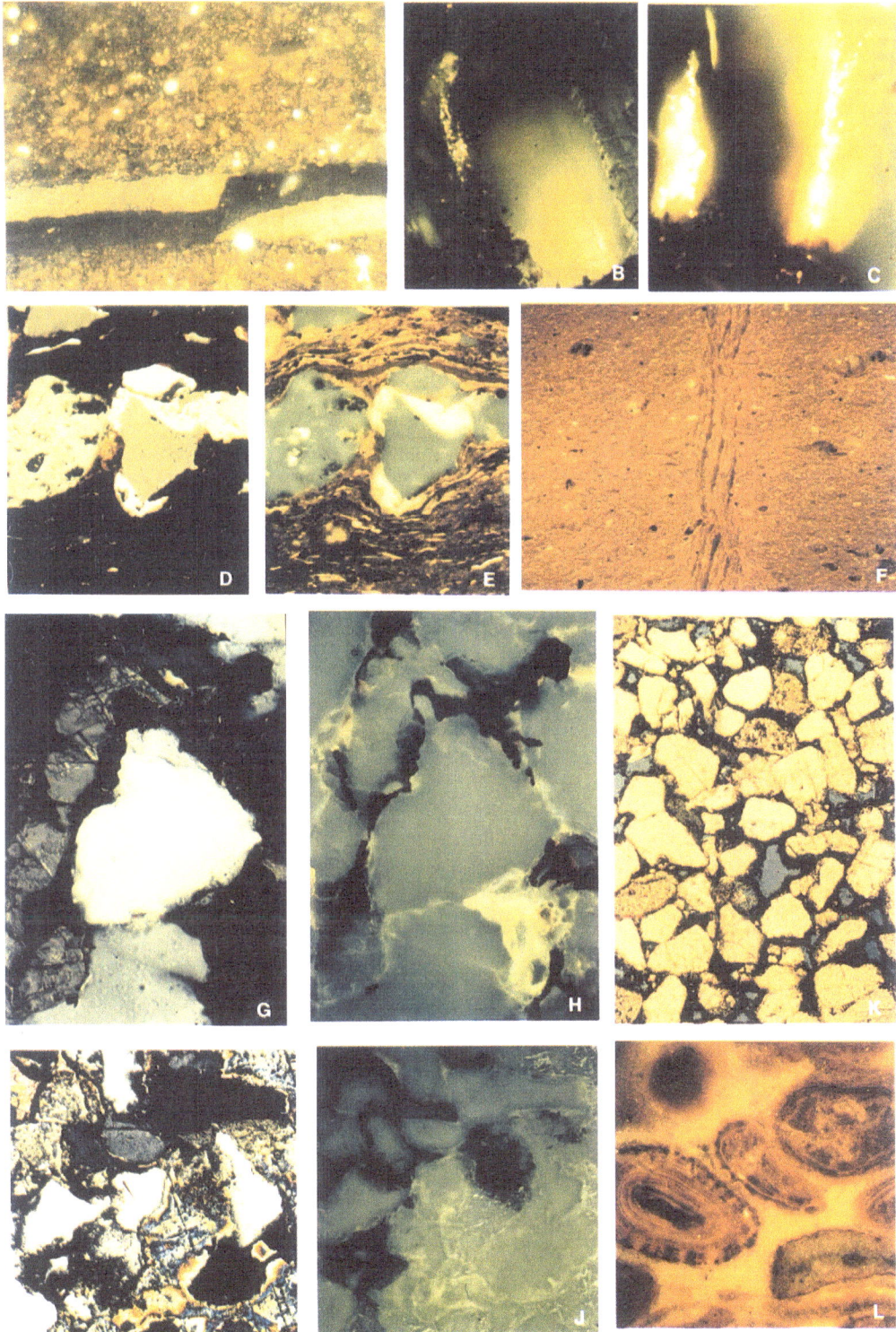

Plate 2.   A) Solid bitumen in a partly open horizontal fracture in overmature Posidonia Shale (Haddessen borehole) form the Hils area, northern Germany. Polished surface, reflected white light, 1 cm = 30 µm; **B**) Fluorescing, oil-bearing fluid inclusions in calcite filling a horizontal fracture in mature Posidonia Shale (Harderode borehole) from the Hils area, northern Germany. Polished surface, fluorescence mode, 1 cm = 75 µm; **C**) Fluorescing, oil-bearing fluid inclusions in calcite in mature Posidonia Shale (Harderode borehole) from the Hils area, northern Germany. The inclusions seem to have migrated from the Posidonia Shale (lower dark part of the picture) into the calcite along growth fronts or fractures. Polished surface, fluorescence mode, 1 cm = 30 µm; **D**) Quartz grains surrounded by a rim of cement in Kimmeridge Clay from the Brae area, North Sea. Polished thin section, 1 cm = 30 µm; **E**) Same view as D, but in an incident light fluorescence mode. Note the fluorescence of the cement indicating a partly organic nature. Other yellow fluorescing particles are alginites; **F**) Fracture perpendicular to bedding in Brown Limestone, Egypt. Polished surface, fluorescence mode, 1 cm = 150 µm; **G**) Primary quartz, diagenetic calcite and solid bitumen (dark) in a Rotliegend sandstone, northern Germany. Polished thin section, 1 cm = 30 µm; **H**) Same view as G, but in an incident light fluorescence mode. The dark solid bitumen can easily be recognized; **I**) Anhydrite cement and solid bitumen surrounding primary quartz grains and illite overgrowth cement in a Rotliegend sandstone, northern Germany. Polished thin section, 1 cm = 30 µm; **J**) Same view as I, but in an incident light fluorescence mode. The section cemented by dark solid bitumen can be easily differentiated from the area cemented by anhydrite; **L**) Ooids in a micritic groundmass in Asphaltkalk from the Hils area, northern Germany. The fluorescence is due to the presence of disseminated bitumen impregnating ooids and groundmass. Polished surface, fluorescence mode, 1 cm = 150 µm.

6.2   Effects of petroleum impregnation in reservoir rocks

Macroscopic, microscopic and geochemical features of reservoir rock impregnation were studied on Lower Permian (Rotliegend) reservoir rocks in northern Germany (Brauckmann and Littke, 1989). These fine-medium grained sandstones were cored in several wells between the Elbe and Weser rivers at depths between 2500 and 5000m and are partly gas-productive. They are generally several metres or a few tens of metres thick and interlayered with mudstones (Fig. 55). The latter contain extremely little organic matter (usually less than 0.2% C_{org}) and are not regarded as potential or former petroleum source rocks (see Chapter 5.1).

The sandstones are composed of quartz and variable percentages of rock fragments, feldspars and mica as primary components. Macroscopically, a peculiar mottling is their major characteristic, i.e., the colour varies from light-gray to dark-gray in an irregular way. This colour change seems to be independent from the primary composition of the sandstones, e.g., there is no obvious relation between grain size and colour or between primary mineralogic composition and colour. From the study of the relatively small cores it remains unclear, whether the dark areas belong to isolated dark volumes in these rocks or whether they are interconnected by dark channels. Consistently dark, thick (> 1cm) layers were not observed.

Microscopical observations were performed in transmitted light, in a fluorescence (incident light excitation) mode, and in reflected light. The last two methods revealed that the dark staining of the sandstones is due to the presence of thin (usually less than 5µm) organic "skins" around quartz grains and other primary components (Plate 2, G and H). Combined transmitted light and fluorescent light studies revealed that organic matter surrounds authigenic illite cements which occur as overgrowth on quartz grains. As the illites are about 190 million years old according to K/Ar-dating, a post-Triassic emplacement of the organic matter is concluded. The coatings, according to their reflectance (~ 1.2-2.5%), can be classified as im?onite, viz as a high-reflecting solid bitumen. The differences in reflectance are explained by regional variations in temperature history and maturation. Solid bitumen reflectivities can be converted into vitrinite reflectance values according to the correlation of Jacob (1989) and are in the reflectance range of 1.2 to 2% roughly similar to the latter. Thus, all samples are at present at maturity levels corresponding to the end of oil generation or to the overmature stage of gas generation (see Figs. 41 and 42).

55

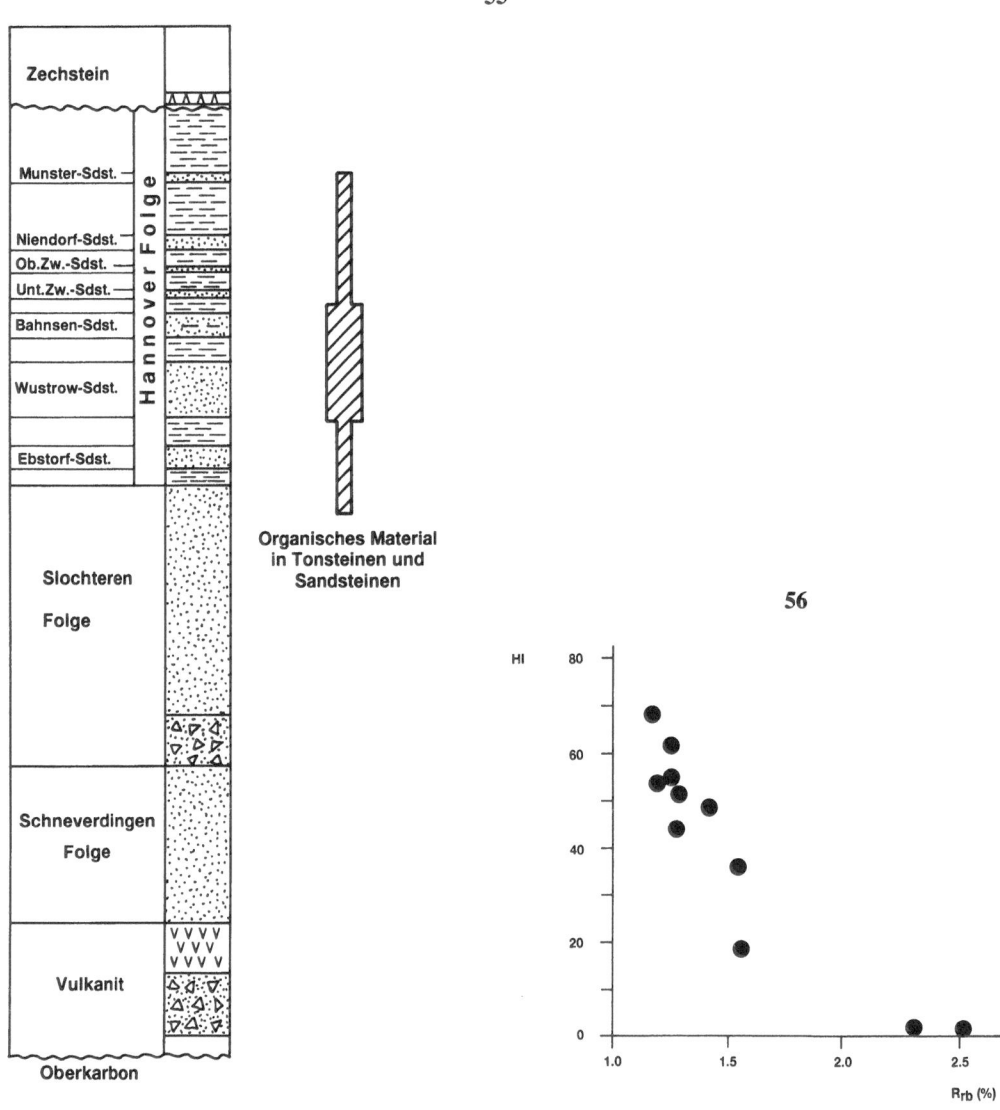

Organisches Material
in Tonsteinen und
Sandsteinen

56

Fig. 55:   Schematic lithologic column through the Lower Permian in northern Germany (area
             between Elbe and Weser rivers).

Fig. 56:   Plot of solid bitumen reflectivities versus HI-values from Rock-Eval pyrolysis (see
             Chapter 2) for samples from Rotliegend sandstones, northern Germany. Note that
             accurate reflectance measurements could only be performed on few of the studied
             samples, because the solid bitumen particles are often very thin; therefore kerogen
             concentration (Fig. 1) was necessary as a preparative step, before measurements could
             be performed.

The solid bitumen is regarded as the relic of a former oil-filling (see Curiale, 1986) of the pores in the sandstones of the Lower Permian which took place, when the sandstones were at temperature and maturity levels much below the present ones. Based on results from basin modelling which indicate petroleum generation from the Carboniferous during Jurassic times, it is assumed that the oil-impregnation took place "shortly" after the formation of the authigenic illite (Brauckmann and Littke, 1989). Further subsidence of the reservoir sandstones and temperature increase may have led to a cracking of the formerly present oil into gas and solid bitumen. This interpretation is supported by the burial history of these sandstones, but other explanations such as chemical precipitation of asphaltenic compounds are not excluded.

In this context, it is interesting to know whether the gas that presently fills some of the sandstones is derived (partly) from the cracking of the former oil or whether it was provided later by a different source. The first possibility would imply that (part of) the hydrocarbons are in place already since the Jurassic. A key towards an answer for this question is the comparison of the maturity measured on the solid organic matter (solid bitumen reflectivity, T_{max}) on one hand and on mobile products (molecular maturity parameters) on the other. First results at least for one well indicate that maturity measured on solid organic matter is lower than that measured on aromatic hydrocarbons. This implies that (part of) the mobile products which presently fill the sandstones are derived from a deeper, more mature source and that oil-cracking in place is not the (only) generative mechanism.

Besides the fact that the formation of the solid bitumen provided hydrocarbon gas, it also influenced the porosity development to a great extent. In general, the dark parts of the mottled sandstones are characterized by the thin solid bitumen coatings, whereas organic particles are absent in adjacent light gray (almost white) parts of the sandstones. The latter are in many cases completely cemented by calcite or anhydrite (Plate 2, I and J). In contrast, the pores between the solid bitumen skins are often empty (Plate 2, K). This may indicate that cementation took place at a time, when the dark parts of the sandstones were oil-filled so that cementation was inhibited.

The organic coatings occupy less than 1% of the volume of the sandstones (best microscopic estimate is possible by fluorescence microscopy; see Plate 2, G and H) corresponding to less that 0.7% C_{org}, in most cases about 0.3-0.4% C_{org}. This small percentage of organic matter is sufficient to cause a rather drastic colour difference, because it occurs as a coating on the primary grains in the sandstone. The solid bitumen still contains hydrogen; i.e. HI-values from Rock-Eval pyrolysis range from about 20-100 (see Fig. 2B). HI-values are clearly controlled by maturity as revealed by a plot versus solid bitumen reflectivity (Fig. 56). Only the less mature solid bitumen still possesses a small gas generation potential.

The source for the oil and later gas in the sandstones in the Weser-Elbe area is probably the Upper Carboniferous which is known to be rich in organic matter (coal-bearing). This assumption is in accordance with molecular data on bitumen characteristics (not shown here), but no positive proof could be given due to the overall high maturity. Only in the uppermost (Munster) sandstone (see Fig. 55) of the least mature well, molecular data (even/odd predominance; phytane/pristane ratio above 1.0) indicate another source, i.e., a carbonate or evaporite source rock. In the study area, carbonate source rocks are known to occur in the Upper Permian (Zechstein), but not in the Carboniferous.

One important question that remains to be answered is, why the solid bitumen occurs only in patches or channels in the sandstone rather than being homogeneously distributed. A possible interpretation supported by theoretical transport models (oral communication with U. Bayer, Potsdam) is that these channels represent migration pathways for oil, which were occupied when two phases (oil and water) were present in the sandstone. In this case, the oil seems to flow through specific channels rather than to fill out the pore space homogeneously (England et al., 1987).

Another example of a petroleum reservoir on which microscopic and geochemical studies were performed is the Asphaltkalk of the Hils syncline in northern Germany. This oolithic limestone is exposed close to the surface. Long-lasting subsurface mining for the asphaltic rock gave access to the interior of this petroleum reservoir. Organic carbon percentages reach 5% (Horsfield et al., 1991) and are higher close to major fractures or in structural higher parts of the different members of this reservoir (see Heckers, in prep., for details). Geochemically, the organic matter is almost completely composed of bitumen, i.e., more than 90% of the organic matter are extractable using non-specific organic solvents (dichloromethane). This indicates clearly that the organic matter is indeed of allochthonous origin (migrated oil) rather than composed of primary, plant-derived particles. The latter are extremely scarce in the Asphaltkalk based on microscopical studies.

Only very little organic matter is visible as microscopically identifiable particles in the Asphaltkalk. Nevertheless, optical studies provide important clues towards a resolution of the location of the organic matter, if extracted and non-extracted rock pieces are compared. In non-extracted rocks, both ooids and other calcareous particles as well as the micritic groundmass exhibit strong yellow fluorescence (Plate 2, L) which changes after extraction (see Horsfield et al., 1991 for details). It should be noted that the Asphaltkalk appears microscopically dense (with the exception of few fractures) at present; therefore the mode of impregnation is not clear. It possibly occurred early in the diagenetic history of the Asphaltkalk, probably at the time of intrusion of the "Vlotho massif" (see Chapter 5.5). Due to Mid-Cretaceous uplift close to the surface, the oil that once occupied the pore space in the Asphaltkalk was converted into an asphaltic residue by microbial activity.

In summary, optical methods if combined with geochemical analyses provide an important tool in reservoir characterization, especially if reservoir-filling oil was converted into solid residues by temperature increase, chemical reactions or uplift combined with bacterial degradation.

7.   MIGRATION OF OIL AND GAS IN COALS

7.1   Overview

Petroleum migration in highly permeable carrier beds and reservoir rocks is generally regarded as a buoyoncy-driven process called secondary migration. In contrast, the mechanisms which cause the expulsion of petroleum from source rocks are less well understood. Difficulties in understanding of this "primary migration" arise, because most source rocks are characterized by low permeabilities. Nevertheless oil molecules which are larger than pore throats are able to escape form these sedimentary rocks. Possible mechanisms were extensively discussed, e.g., by Durand (1988). As coals are known to be among the least permeable of all source rocks, the understanding of expulsion of oil and gas from coals is even more difficult than form clastic and carbonate source rocks.

Primary migration in source rocks follows petroleum (= oil and gas) generation, which in turn is a function of rank and kerogen composition (Hunt, 1979; Tissot and Welte, 1984). Whether petroleum generation in coals is an economically important process was controversially discussed in the past (e.g., Katz et al., 1991; Durand and Paratte, 1983; Hunt, 1979, 1991). Six groups of observations strongly support the assumption that great masses of petroleum are generated in coals:

-       Liptinites constitute generally between 5 and 20% of humic coals. These petrographic constituents (macerals) are known to be hydrogen-rich (van Krevelen, 1961) and act as source of petroleum. In many conventional petroleum source rocks, liptinites also constitute between 5 and 20% of the rock (by volume), but are surrounded by a mineral groundmass rather than by an organic, mainly vitrinitic groundmass. It should be noted, however, that liptinites in coals are derived from other precursors (e.g., spores, pollen, resins, cuticles) than liptinites in clastic and carbonate source rocks and will generate different products upon maturation (Given, 1984).

-       Upon artificial maturation (pyrolysis), most coals generate significant quantities of petroleum compounds. The gas/oil ratio and the ratio of aromatic over n-alkyl moieties are often greater in coal pyrolysates than in kerogen pyrolysates derived from marine or lacustrine source rocks (Larter and Senftle, 1985; Horsfield, 1989; Katz et al., 1991; Powell et al., 1991). An example of a pyrolysate from a pure, handpicked vitrinite from a Carboniferous coal seam is shown in Fig. 57. Aromatic hydrocarbons such as benzene, toluene, and xylene are among the major pyrolysis products, but also long chain hydrocarbons (n-alkanes and n-alkenes) are generated. The latter are typical products of oil source rocks.

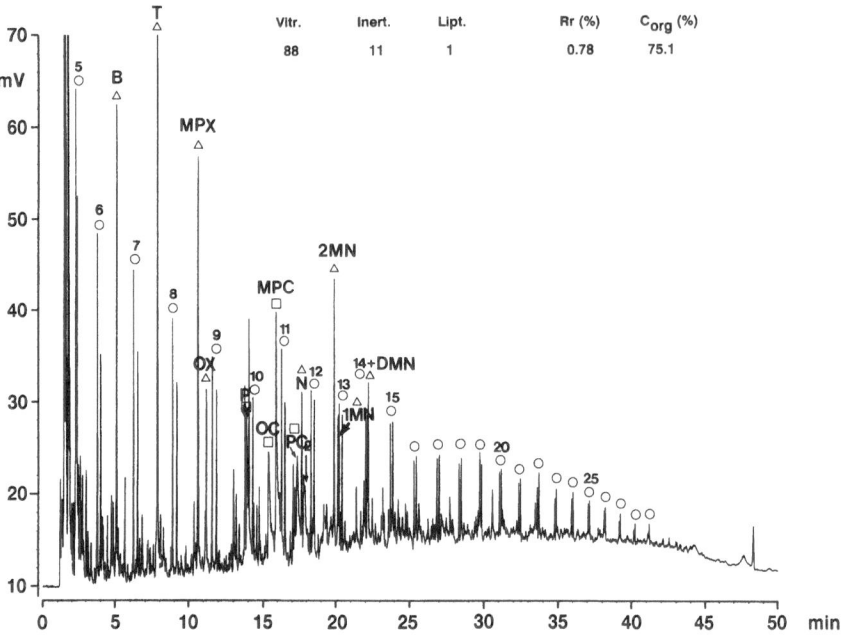

Fig. 57: Pyrolysis-gas chromatogram of a vitrinite-rich coal sample from the Ruhr area, western Germany (from Littke and Leythaeuser, in press). Long-chain hydrocarbons (*n*-alkane/*n*-alkene doublets) were produced from this material, which are thought to indicate an oil-generation potential of the sample (Horsfield, 1989). Note that not all but most vitrinite-rich samples from the Ruhr area produced these compounds and that vitrinite-rich kerogens from adjacent siltstones produced in most cases less long-chain hydrocarbons than vitrinite-rich coal samples from the same depth interval. Abbreviations: B=benzene, T=toluene, MPX=meta- and paraxylene, OX=orthoxylene, P=phenol, OC=orthocrysol, MPC=meta- and paracrysol, N=naphthalene

- Bituminous coals contain great quantities of bitumen that can be released by solvent extraction (Rice et al., 1989). This bitumen consists mainly of heterocompounds (organic N,S,O-compounds) and asphaltenic molecules. The hydrocarbon percentage of the extractable organic matter is lower than in most clastic and carbonate source rocks, but still in the range of about 10 - 25% (Littke et al., 1990).

- Coals contain great volumes of hydrocarbon gases, mainly methane (Jüntgen and Karweil, 1966a,b), which is under favorable conditions even producible from the coal beds themselves (Rightmire et al., 1984) after water is removed from the cleat system.

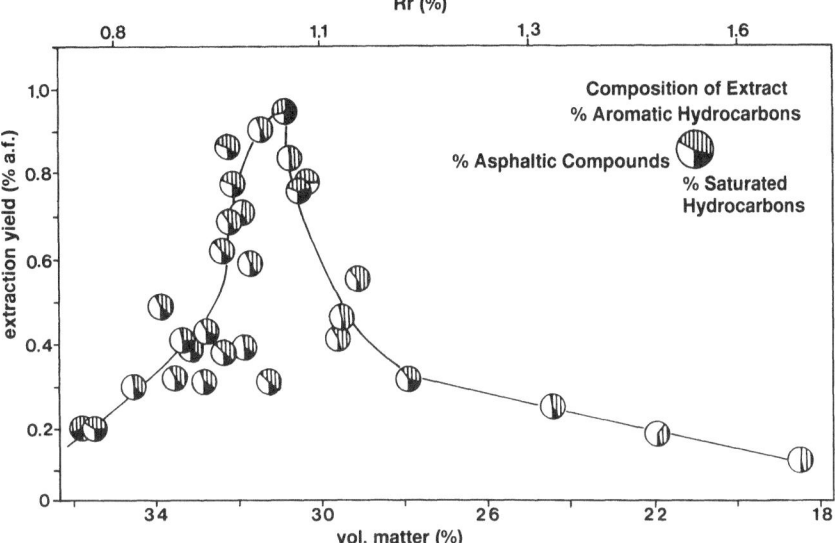

Fig. 58:   Extraction yields (% ash-free) and composition of extracts in relation to the volatile matter yield and vitrinite reflectance of Carboniferous coals from the Ruhr area (highly mature coals) and from the Saar district, Germany (from Littke and Leythaeuser, in press; after Leythaeuser and Welte, 1969). Data suggest the occurrence of an "oil window" in very much the same way as well established for clastic petroleum source rocks (Tissot et al. 1971).

- Coal-bearing strata are often associated with commercial gas fields, e.g., in North America (Masters, 1984, Rice et al., 1989) and western Europe (Lutz et al., 1975). During the last decade, an increasing number of oil accumulations were also attributed to coal source beds (e.g., Thomas, 1982; Hvoslef et al., 1988; Noble et al., 1991).

- The fact that coals originate from land-plant derived organic matter, which is transformed into peats and grades upon maturation into anthracite and finally into graphite implies a significant loss of volatile products. The latter two observations indicate some petroleum generation in coals and some petroleum expulsion from coals.

Nevertheless, doubts about the petroleum generation potential of coals are raised due to average chemical properties. For example, atomic hydrogen/carbon ratios are often smaller than 1 (Powell et al., 1991), whereas ratios greater than 1.2 are typical for oil and gas and for organic matter in marine source rocks. According to hydrogen/carbon and oxygen/carbon ratios, coals are mostly classified as type III kerogens, whereas many marine and lacustrine kerogens are grouped as type II and I kerogens, with higher hydrogen/carbon and lower oxygen/carbon ratios (Tissot and Welte, 1984). This classification implies a low petroleum generation potential of coals (e.g., Hunt, 1979: 274).

However, on a basin-wide scale, not only the hydrogen content of the organic matter but also the total amount of organic material is of crucial importance for the petroleum quantity. A new classification scheme, the source prolificity index (SPI; Demaison and Huizinga, 1991), takes this into account and avoids the normalization to organic carbon. It predicts high petroleum generation capacities for many coal-bearing basins which contain only poorly productive type III kerogen according to hydrogen/carbon ratios or HI-values from Rock-Eval pyrolysis (Table 25).

Name	Age	Thickness (m)	C_{org} (%)	HI (mg hc/g C_{org})	SPI (t hc/m^2)
Coals, Ruhr area	Carboniferous	100[1]	80.0	100-200[2]	16.8[3]
Silt- and Sandstones Ruhr area	Carboniferous	2900[1]	1.0[4]	50-100[2]	7.0
Posidonia Shale	Liassic	30	10.6	663	4.3

Table 25:   Stratigraphic age, thickness, mean organic carbon content, typical hydrogen index values, and source prolificity index values (SPI; Demaison and Huizinga, 1991) for different organic matter-rich rocks and different basins, respectively.

[1] Cumulative thickness of all coal seams and all clastic rocks, respectively.
[2] Higher values are valid only for the upper part of the section (Littke et al., 1989).
[3] Based on the assumption of an average specific weight of 1.4 t/m^3
[4] see Scheidt and Littke (1989)

Geochemical studies of the effects of generation of petroleum hydrocarbons in coals are abundant and only a brief review is given here. Brooks and Smith (1967), Leythaeuser and Welte (1969), and Hollerbach and Hagemann (1981) documented systematic changes in soluble organic matter yields and compositions (especially those of long-chain n-alkanes) as a function of coal rank. As an example, Fig. 58 shows progressive soluble organic matter yield increases and gross compositional changes maximizing around a rank level equivalent of 30% volatile matter ($\approx 1.2\% \ R_r$) for Upper Carboniferous-age coals from the Saar area, Federal Republic of Germany. This change and the subsequent trend of decreasing soluble organic matter contents with further rank increase is equivalent to the "petroleum generation curve," which was established later for shale source rocks (Tissot et al., 1971). Furthermore, rank-dependent changes in the molecular composition of extractable hydrocarbons were found for a series of Westfalian coals from the

Ruhr area, Germany (Radke et al., 1980, 1982, 1984), which clearly indicate that petroleum-type compounds are generated in coals with coalification progress.

During the last decade, several authors suggested that high proportions of liptinite in general or of specific liptinite macerals favor effective oil generation in coals (Snowdon and Powell, 1982; Risk and Rhoades, 1985; Thompson et al., 1985; Khavari-Khorasani, 1987; Horsfield et al., 1988), although other macerals such as fluorescing vitrinite (Horsfield et al., 1988) or detrovitrinite (Powell et al., 1991) are also considered as source for oil. On the other hand, Durand and Paratte (1983) studied coal samples from various basins and concluded that coals in general have a fairly good potential for oil generation.

In summary, the petroleum (oil and gas) generation capacity of many coals cannot be disputed. There seems to be some agreement among coal researchers that the oil generation potential of coals should be studied using petrographic and geochemical techniques, among which pyrolysis-gas chromatography is of crucial importance (e.g. Powell et al., 1991). It remains to be answered whether generated oil can be expelled from the coals or whether it is trapped therein until high maturation stages are reached, at which point oil to gas cracking takes place.

In the following, a review of problems related to oil and gas migration within coals and to expulsion from coals will be given which is based on a recent publication (Littke and Leythaeuser, in press). The first sections will provide background information on porosity and permeability of coals ("Physical Conditions of Migration") and on concepts for migration of oil in coals ("Physical Migration Mechanisms and Avenues"). Thereafter, "Geochemical and Microscopic Effects of Migration" and "Mass Balance Approaches" are presented and discussed which should provide quantitative information on expelled petroleum masses and qualitative assessment of its molecular composition. Finally, "Migration of Gas" will be briefly treated.

7.2   Physical conditions of migration

Not only petroleum generation, but also the existence of migration pathways is a necessary prerequisite for petroleum migration to occur. If expulsion occurs as a separate-phase flow (Jones, 1980) and not as a diffusive process through the organic matter network (Stainforth and Reinders, 1990), the pore system is of crucial importance for the movement of any bitumen in a source rock. The porosity and permeability of coals are changed by maturation and pressure, and both influence the storage capacity for bitumen, including methane, and the rate of migration. The following discussion aims to review some basic information on this subject.

The pore sizes of bituminous and subbituminous coals as measured by mercury injection range from 5 to more than 100 nm; the bulk being smaller than 7 nm (Thimons and Kissell, 1973). However, the polymer structure of coal renders the conventional porosity concept questionable because i) pores of less than 10nm are in the range of organic molecules and can be interpreted as intramolecular interstices within cross-linked polymers and ii) the phenomenon of solvent swelling indicates that coal pore sizes are a function of temperature conditions.

Pores smaller than 5 nm (= micropores) cannot be measured accurately, but it is estimated that they account for 95% of the internal coal surface and for most of the porosity (Thimons and Kissell, 1973) and that, accordingly, most of the bitumen (including methane) will be present in the micropores. Data by Debelak and Schrodt (1979) indicate that most coal pores are in the range of 0.6 - 2.0 nm. Interestingly, Harris and Yust (1976) report from TEM-observations that liptinites mainly contain mesopores (up to 50 nm) but few or no micropores. They suggested that micropores may be restricted to vitrinitic material.

The total porosity or the open pore volume (Mahajan and Walker, 1978), i.e., the volume percentage of the sum of all pores with diameters between 0.2 nm (diameter of helium) and 50 μm, ranges for bituminous coal between 3 and 18%, corresponding to pore volumes of $0.02 - 0.16$ cm^3/g (Janowsky, 1984) and is greatly affected by maturation (Fig. 59). From the high volatile bituminous coal stage ($R_r \sim 0.7\%$) to the medium-low volatile bituminous coal stage ($R_r \sim 1.4\%$) porosity is decreasing. Towards higher degrees of maturation corresponding to the anthracite stage ($R_r \sim 2\%$), porosity is again increasing (Mahajan and Walker, 1978). The minimum porosity occurs at the end of the "oil generation window" and is due to a loss of microporosity (Janowsky, 1984). This phenomenon is tentatively explained by trapping of generated bitumen within the micropores. This assumption is in agreement with the finding of Harris and Petersen (1979) that solvent extraction creates new micropores in coal. It should be noted that different methods of porosity measurement reveal a similar trend, but that absolute values are vastly different (Fig. 59). Major problems in obtaining accurate information on micropore size distribution relate to the lack of knowledge on the pore shapes and to uncertainties about the correct way of conversion of adsorption data into pore-volume data (Debelak and Schrodt, 1979).

The water saturation of coal pores decreases with increasing maturation, as indicated by moisture contents generally decreasing from about 10% in the high volatile bituminous coal stage ($\sim 0.7\%$ R_r) to 1% in the medium volatile bituminous coal stage ($\sim 1.4\%$ R_r; see Stach et al., 1982). Water in coal pores reduces the mobility of methane (and other petroleum compounds) as well as the adsorption capacity of coal for methane (Janowsky, 1984). These data on moisture content imply both greater mobility and higher storage capacities for methane at high ranks and coincide with the general observation that the adsorption capacity of coal for methane increases with increasing maturity (Fig. 60; Jüntgen and Karweil, 1966b). The adsorption capacity is, however, exceeded by methane generation at high maturities corresponding to >1% vitrinite reflectance (Jüntgen and Klein, 1975).

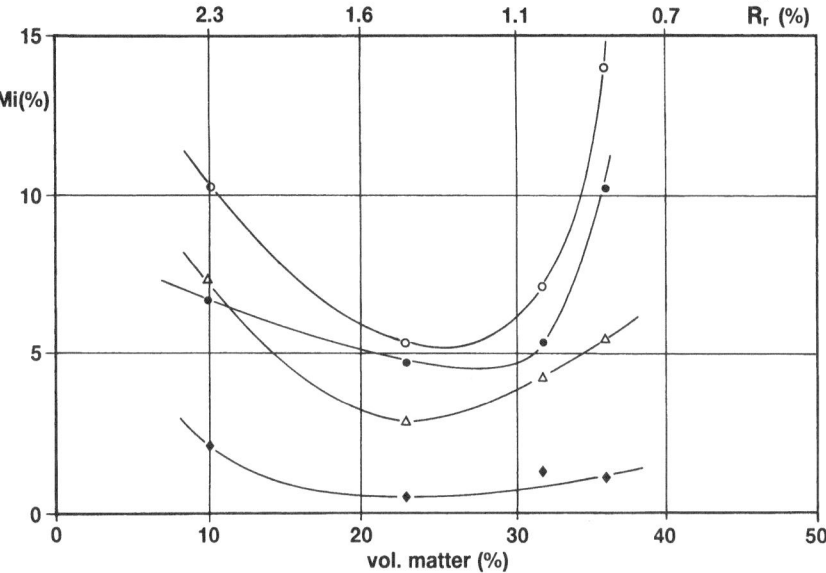

Fig. 59:   Microporosity (γ_{mi}) calculated for coals of variable maturity using different methods (methanol isotherms for pores smaller than 10 nm and smaller than 1 nm, respectively: upper two curves; water isotherms: triangles; bubble pressure method: lowermost curve). A minimum of microporosity is revealed by all methods for bituminous coals with about 20-30% volatile matter (from Littke and Leythaeuser, in press; after Janowsky, 1984).

The permeability of coals greatly depends on the availability of macropores (Thimons and Kissell, 1973; Seewald, 1982) which unfortunately cannot be measured exactly. Macroporosity depends on pressure and cracks and macropores are produced in coals by tectonic uplift, mining or coring. Thus, in situ measurements reveal permeabilities by two orders of magnitude lower than measurements on specimens at the surface (Janowsky, 1984). Jüntgen (1986) compared porosities of coals at great depth and of uplifted coals of similar maturity levels. He found that microporosity is in the same range for both groups of coals, but that larger pores are significantly influenced by uplift (Fig. 61). Coals at great depth show a much lower macroporosity. This implies that porosity measurements on uplifted coals are not accurately predicting porosity at greater depth, i.e., in the depth range where hydrocarbon generation and expulsion is to be expected. As the majority of the bituminous coal deposits in mining areas were uplifted after maximum burial by often more than a thousand meters, most available coal samples are not well suited for prediction of coal porosity in the deep surface.

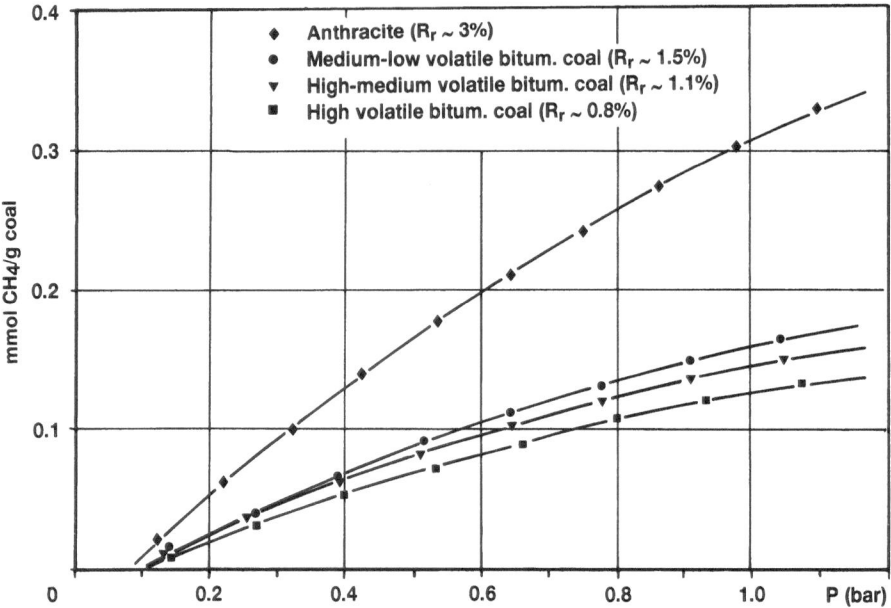

Fig. 60:   Methane adsorption as a function of pressure for coals of variable maturity measured at room temperature. Methane adsorption is increasing both with increasing maturity and with increasing pressure (from Littke and Leythaeuser, in press; after Janowsky, 1984).

Most permeability measurements on coals probably do not accurately predict the permeability for bitumen or methane at depth. Permeabilities for methane were calculated by Thimons and Kissell (1973) for a variety of coals on the basis of Knudsen diffusivity and range between 0.51 and $28 \cdot 10^{-6}$ cm^2/s for unfractured samples. As these calculations only describe permeability in the micropore system, they can be regarded as minimum values and are probably much lower than in in-situ coals (Thimons and Kissell, 1973).

In summary, coal porosity is characterized by the predominance of micropores, which in the absence of fractures allow slow migration only. At low levels of maturation, water saturation of the coal pore system is high and further inhibits hydrocarbon movement. Whether at depth a significant macroporosity increases the permeability is unknown.

7.3   Potential migration mechanisms and avenues

As described in the previous chapter, coals are characterized by small pores which cannot be penetrated easily by petroleum, especially by oil. Many organic molecules such as asphaltenes possess diameters which are in the range of pore throats in coal or even exceed them. The resultant inhibition of any movement of volatile compounds is further enhanced by the presence of water, which occurs in coals in high concentrations mainly at low maturity levels (lignite and subbituminous coal stage).

Nevertheless, theoretical concepts exist which predict that oil expulsion from coals is occurring to a significant extent. Three of these concepts are discussed briefly here. The first concept of oil expulsion from coals occurring in gaseous solution as an effective possible migration mechanism relies on two basic arguments: i) coals are known to be prolific gas source rocks, i.e., over a wide coalification range there is a higher abundance of gas as compared to oil in the coal pore system, and ii) the analogy to type III kerogen-bearing source rocks, where this transport mechanism is suggested to play an important role (Price, 1989). There is geochemical and experimental evidence in support for the latter conclusion. Especially, the composition of high molecular weight n-alkane mixtures expelled from source rocks bearing abundant land plant-derived organic matter (type III kerogen) reveal signs of compositional fractionation according to molecular chain length. This effect is attributed to the process of expulsion in gaseous solution, since gaseous solubilities of long-chain n-alkanes are known to decrease with increasing molecular chain length (Price, 1989). Since the organic matter in both sediments, type III kerogen-bearing shales and coals is derived from similar residues of land plant vegetation, and since there are transitional sediments between both end members ("Brandschiefer," carbonaceous (coaly) shales), it is suggested that this particular transport mechanism will apply in both cases. However, no geochemical evidence was presented so far to indicate that coals have indeed expelled petroleum in gaseous solution. Furthermore, if sizes of organic molecules are in the range of pore throats in coal, even transport in a gas phase cannot mobilize these molecules.

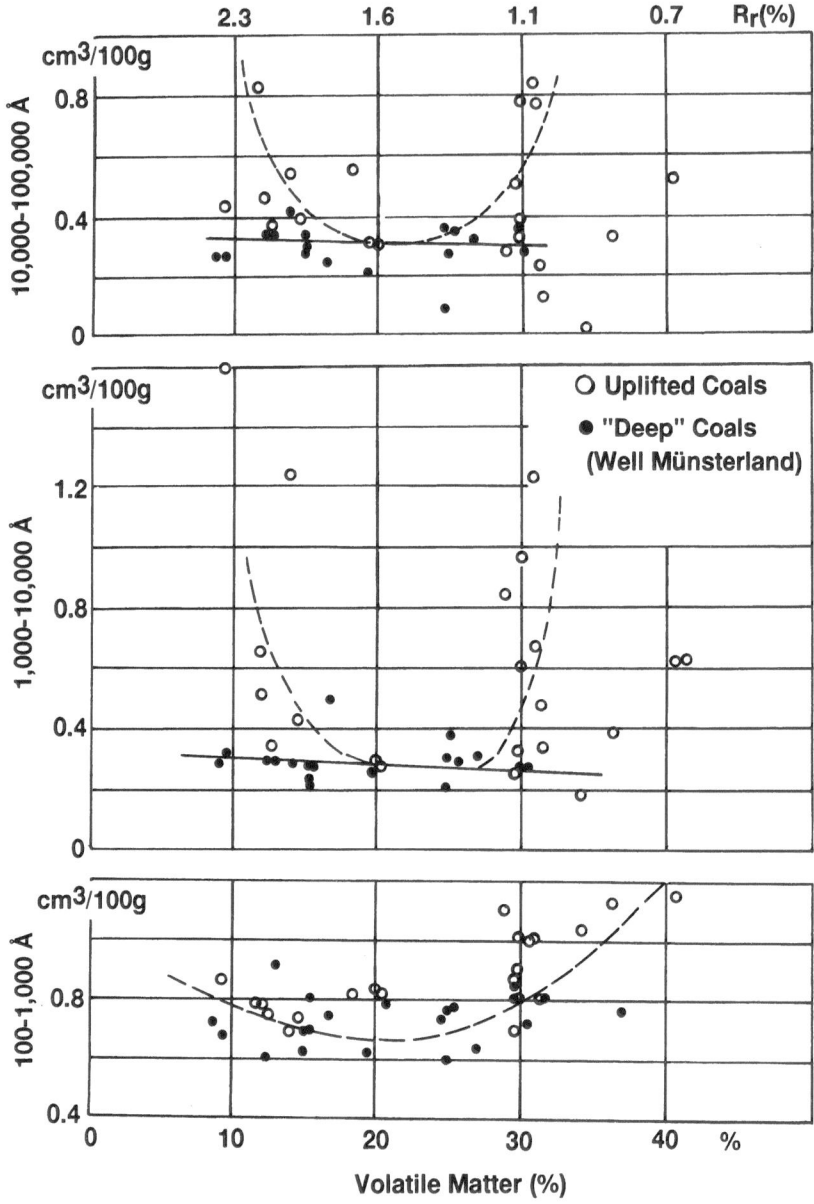

Fig. 61:   Pore volume plotted versus maturity parameters of "deep" coals from well Münsterland (closed circles) and uplifted coals from the Ruhr area mining district for three different pore size classes. Volumes of large pores differ considerably between the two sets of samples at a given maturity level (from Littke and Leythaeuser, in press; after Jüntgen, 1986).

As a second approach to explain petroleum expulsion from coals, it can be assumed that oil is expelled from coals as a separate phase (diphasic flow of oil and water) as from other more conventional source rocks. In this case, capillary pressures become an obstacle for petroleum expulsion. Capillary pressures depend on the presence of at least two fluid phases (petroleum and water) and one solid phase. Durand and Paratte (1983) argued that there is much less water present in coals than in other source rocks at the petroleum generation stage and that, therefore, expulsion is easier in coals and that "very early expulsion flow should be monophasic and capillary pressures inexistent". An early expulsion of oil from coals seems to be indicated by the low amount of thermal extract (S_1-peak of Rock-Eval pyrolysis released at 300°C) and of solvent extract as argued by Durand and Paratte (1983) and Huc et al. (1986). It should, however, be noted that the low extract content of coals is partly an effect of its usual normalization to organic carbon content (which is very high in coals). Furthermore, long lasting extraction produces much greater masses of solvent extracts than Rock-Eval pyrolysis (Table 26). This observation can be explained by a concept of open and closed pores (Radke et al., 1990). Possibly, bitumen confined to closed pores is not released as thermal extract; this is an alternative explanation of the notoriously low S_1-peaks of coals.

A third possible concept in favour of an effective petroleum expulsion from coals is the interaction of the micropore system (in which slow diffusion takes place) with a macropore system. Primary macropores in the lm-range are present in coals in a maceral called fusinite, which is a relic of a higher land plant tissue with cell lumina which are either empty or filled with carbonates, sulfides, or exsudatinite. Exsudatinite was defined by Teichmüller (1974) and was interpreted as a relic of a former oil in coal pores. The amount of the pores in fusinite can reach more than one percent of the total coal volume, and many of these are filled with exsudatinite at maturity levels of about 0.9-1.0% R_r e.g. in coals from western Germany (Littke and ten Haven, 1989). In contrast, fractures (cleats) in the same coals are only rarely filled by exsudatinite. The macroporosity in fusinite is an intraparticle porosity in which oil may be stored, but from which it cannot be expelled. The lack of exsudatinite in fractures seems to indicate that these macropores were not available at the time of petroleum generation, i.e., they may be a product of later uplift (see previous section). Another possible explanation for the above phenomenon is, however, that petroleum could migrate along these fractures out of the coals, and that exsudatinite was only formed where petroleum was trapped for a long time in the coals, such as in the case of the fusinite pores.

In summary, concepts allowing for an expulsion of oil from low-permeable coals exist and include solution of oil in gas, reduced capillary pressures due to low water contents, and availability of fractures. There is, however, no positive proof that one of these concepts is valid for natural expulsion of oil from coals.

Coal Seam	V	R	O	J	C
Depth (m)	1198	1300	1382	1429	1529
R_r (%)	0.70	0.75	0.78	0.81	0.92
R_c (%)	0.83	0.86	0.83	0.92	0.91
T_{max} (°C)	428	431	433	435	439
HI (mg/g C_{org})	234	225	207	217	209
S_1 (mg/g C_{org})	4	4	5	8	10
Max. C_{org} (%)	77	80	84	81	86
Max. extract yield (ppm)	65,600	75,600	81,000	42,600	34,200
Extract yield (mg/g C_{org})	76	71	69	47	39
Aliph. (% of extract)	3.25	4.8	6.2	6.8	8.0
Arom. (% of extract)	8.0	10.4	9.9	13.0	14.8
CPI_{27}	1.32	1.19	1.13	1.07	1.00
Pri./Phy.	7.3	7.0	5.8	5.3	4.6

Table 26:   Average maturity parameters for coals of five depth intervals from well Nesberg 1 (after Littke et al., 1990; see Fig. 63). R_r: vitrinite reflectance measured in oil immersion. R_c: vitrinite reflectance calculated on the base of MPI1 (Radke et al., 1982); T_{max}: temperature of maximum pyrolysis yield; S_1: thermal extract yield of Rock-Eval pyrolysis; HI: hydrogen index; Aliph.: aliphatic hydrocarbons; Arom.: aromatic hydrocarbons; CPI_{27}: Carbon Preference Index; and Pri./Phy.: pristane/phytane concentration ratio. Max.: maximum values (all other data are mean values).

7.4   Geochemical and microscopic effects of petroleum migration

A microscopic observation (besides the occurrence of exsudatinite) thought to be related to petroleum generation and migration is the increase in abundance of fluorescing vitrinites within the zone of oil generation (Kalkreuth et al., 1991), i.e., at the maturation stage at which maximum yields of soluble organic matter are extractable from coals (Fig. 62; Radke et al., 1980). It was hypothesized that petroleum-like material generated from liptinites has migrated over short distances and impregnated adjacent vitrinite grains (Teichmüller and Teichmüller, 1982).

Petroleum migration in coal seams is supported by a comparison of extract compositions of different coal lithotypes at the same maturation stage. For instance, Allan and Douglas (1977) extracted concentrates of vitrinites and sporinites obtained from the same coal seam. Although these macerals differ greatly in their H/C ratios and structure (see van Krevelen, 1961 and Given, 1984), the *n*-alkane distribution of both macerals was very similar. This can most easily be explained by a redistribution of *n*-alkanes within the coal seams leading to a homogenization in n-alkane composition. This conclusion was confirmed by the study of handpicked, lithologically homogeneous specimens from coal seams (Littke et al., 1990; Fig. 63). Though the maceral composition of these samples is highly variable, the extract composition of samples from a single coal seam remains quite uniform. In Fig. 64a-c, liptinite over liptinite + vitrinite ratios as a measure of maceral composition are plotted versus three geochemical parameters. Obviously, the ratio of saturated over saturated + aromatic hydrocarbons is poorly related to maceral composition (Fig. 64a). Powell et al. (1991) also reported poor correlations between liptinite content and pyrolytically released C_{15+}-normal hydrocarbons. Almost not influenced by maceral composition is the ratio of nonacosane (n-C_{29} alkane) over nonacosane + nonadecane (n-C_{19} alkane; Fig. 64b). It should, however, be noted that there is a considerable difference between two adjacent, lithologically different samples from the uppermost coal seam (seam V, see arrows in Fig. 64b). Also, there is a significant shift towards short-chained n-alkanes in the most mature coal seam (seam C, 0.92% R_r). The composition of aromatic hydrocarbons, exemplified by the ratio of methylphenanthrenes over methylphenanthrenes + trimethylnaphthalenes, is obviously not related to maceral composition (Fig. 64c). Radke et al. (1990) concluded that aromatic hydrocarbons, which are accessible to extraction with nonspecific solvents, "do not generally correlate with maceral composition", "represent products of early thermal evolution" and are mobile within the coals. Thus, there is in general little or no correlation between extract composition on one hand and maceral composition of coals on the other.

A major exception is a pair of bright and dull coals from coal seam V at the lowest maturity level shown in Fig. 64 (high volatile bituminous B coal, 0.70% R_r). The dull coal is liptinite-rich and shows a predominance of long-chain over short-chain n-alkanes (Fig. 65, see also Fig. 64b), whereas the vitrinite-rich bright coal is characterized by a predominance of short-chain n-alkanes. In view of the small depth difference of both samples (1198.32 vs. 1198.35 m), this observation is explained by a limited redistribution and migration of saturated hydrocarbons at the 0.7% R_r maturity stage. At higher stages of maturity, a more efficient redistribution is to be assumed, leading to a more uniform extract composition. Another possible explanation for these data is a difference in accessibility of the solvent to the two different coal samples (see Littke et al., 1990).

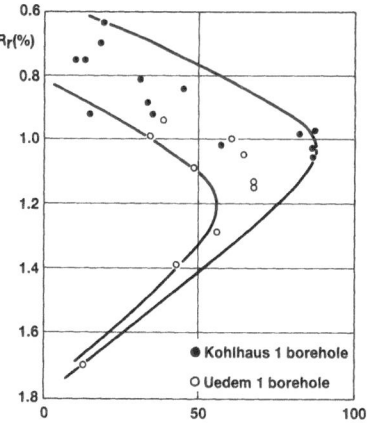

Fluorescent vitrinite (% of total vitrinite)

Fig. 62:   Percentage of fluorescent vitrinite (% of total vitrinite, the most common maceral group in coals) in coals from the Ruhr area, western Germany, plotted versus vitrinite reflectance. At the 1.0-1.1% R_r-level both a maximum in the percentage of fluorescent vitrinite and a maximum in extraction yields (see Fig. 58) was observed. The fluorescence of vitrinite is tentatively suggested to be related to the generation of soluble organic matter (from Littke and Leythaeuser, in press; after Radke et al., 1980).

A key to the recognition of the effects of primary migration in shale source rocks was the combination of detailed geochemical analyses and mass balances for sample suites taken systematically from the center of thick source rock intervals towards their contact with adjacent reservoir sandstones (e.g. Mackenzie et al., 1987; Leythaeuser et al., 1988b, 1988c). A similar approach was also applied to the above discussed coal seam C (Fig. 63) in order to find evidence for or against expulsion of petroleum from coal (Littke et al., 1989). This seam was chosen specifically since it is mature (0.92% R_r; Littke et al., 1989) and it is directly overlain by a sandstone unit, which could have provided a high-porosity/high permeability migration avenue to accommodate the petroleum migrating out of this coal seam. There is microscopic evidence to suggest that the basal part of this sandstone is bitumen-impregnated. The underlying siltstone is much finer-grained and lighter and shows no macroscopic or microscopic signs of impregnation. Throughout the thickness of this coal seam there is no systematic variation of the maceral composition (Fig. 66), i.e., the proportions of vitrinite, inertinite and liptinite macerals vary more or less randomly with depth. In contrast, the trends of C_{org}-normalized n-alkane yields reveal regular depth trends. There is a pronounced decrease of yields of total n-alkanes (C_{15}-C_{30}), as well as of selected normal and isoprenoid hydrocarbons (Fig. 67a, b) from the deepest part towards the shallowest part of this seam. For example, the yield of total n-alkanes decreases from values between 750 and 920 in the center and bottom part to a value of 500 µg/g

C_{org} in the coal sample at 1529.05 m, which is closest to the sandstone contact. These systematic trends of concentration decrease are interpreted to reflect preferential expulsion of C_{15+}-hydrocarbons from that part of the coal seam which is adjacent to the sandstone. The alternative explanation of the regular depth-trends of hydrocarbon concentrations shown in Fig. 63 as a result of a gradual change in the depositional environment of this coals, e.g., from anoxic to oxic conditions, is - based on maceral analysis data (Fig. 66) - considered unlikely. If such an evolution had prevailed, a regular trend towards a predominantly inertinitic maceral composition would be expected from the bottom towards the top of coal seam C. It should be noted that concentrations of aromatic hydrocarbons vary less regularly with depth as previously established for shale source rocks (Leythaeuser et al., 1988b, c).

The concentration and composition of the *n*-alkanes from the two shallowest coal samples is shown in Fig. 68 in comparison with those of the deepest coal sample (1529.05 and 1529.55 m versus 1530.11 m). A marked concentration decrease is observed for the coal samples closest to the sandstone contact, e.g., sample 1529.05 m has only about half the concentration of *n*-alkanes as compared to sample 1530.11 m. These data can be used to calculate relative expulsion efficiencies (Fig. 69) according to previously established concepts (Leythaeuser et al., 1988b), taking the *n*-alkane concentration of the deepest sample 1530.11 m as the reference. If the latter sample has experienced little or no expulsion, then the calculated relative expulsion efficiency values would be close to the absolute values. If not, then the trend of relative molecular expulsion efficiencies shown in Fig. 69 can at least be interpreted to indicate the primary migration mechanism. Three conclusions can be reached from these data: i) molecular expulsion efficiencies increase from around 30% in the center of this coal seam to values between 40 and 50% at the topmost sample, ii) the zone of enhanced petroleum depletion near the sandstone contact is much thinner than that in a shale source rock, e.g., it is about 5 m wide in the Kimmeridge Clay Formation (Leythaeuser et al., 1988b), and iii) the trends of expulsion efficiencies remaining rather uniform over a wide molecular range indicate expulsion of oil as a separate phase fluid. The absence of molecular fractionation effects with expulsion has previously been established as an argument in favour of oil-phase migration (Leythaeuser et al., 1988b).

The geochemical data of coal seam C discussed above represent evidence that coal can indeed expel petroleum hydrocarbons into adjacent carrier beds. The question of whether petroleum expulsion from this coal seam C represents an exceptional case due to the close proximity to an effective carrier bed remains to be answered. The similarity in elemental composition and pyrolysis data of coal seam C with those of coals of the same maturity which are not overlain by sandstone contradicts, however, any drastic differences in expulsion.

150

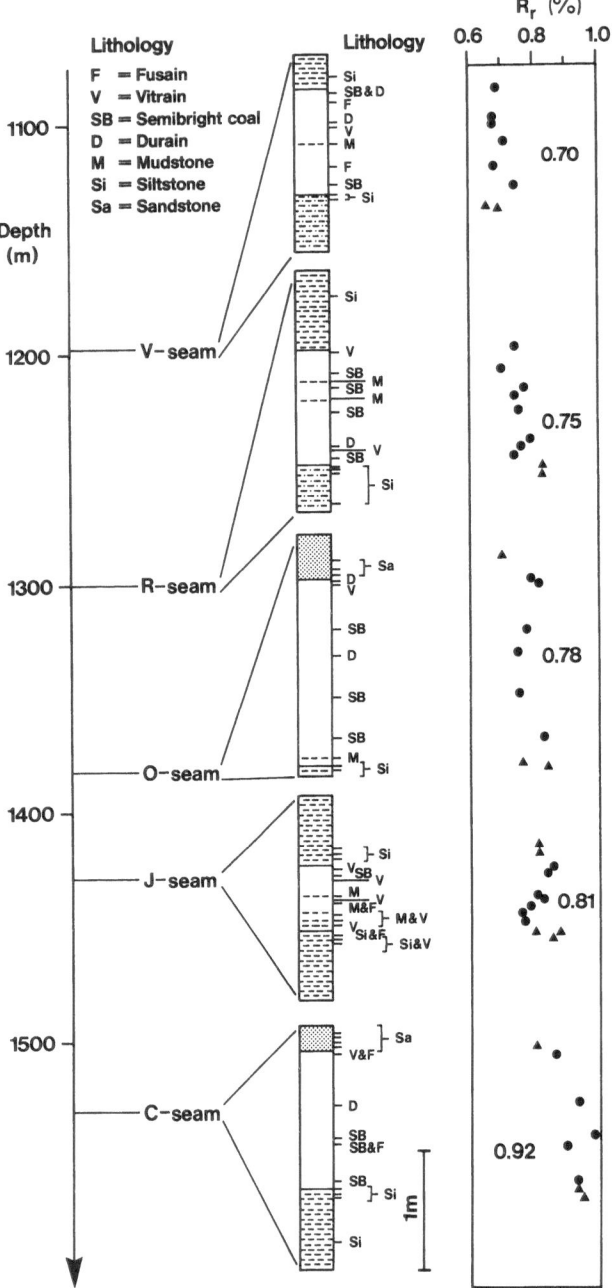

Fig. 63: Depth and lithology of samples from five intervals in borehole Nesberg 1 (after Littke et al., 1990). The column on the right represents vitrinite reflectance (R_r) values for samples from coal seams (circles) and adjacent clastic rocks (triangles). Mean values are given for each interval.

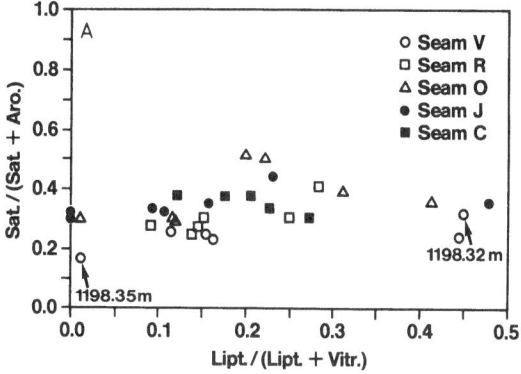

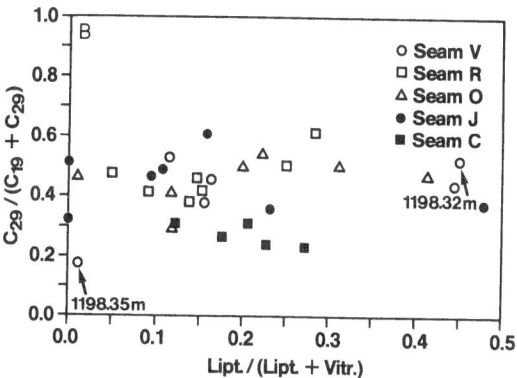

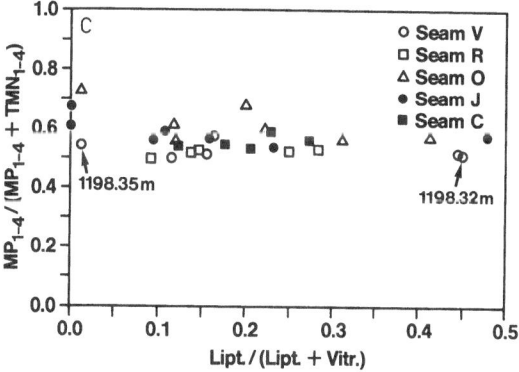

Fig. 64a-c: Liptinite/(liptinite + vitrinite) ratio versus saturated hydrocarbons/(saturated + aromatic hydrocarbons) (a), $C_{29}/(C_{19} + C_{29})$ *n*-alkanes (b) and (c) sum of 1-, 2-, 3-, 9-methylphenanthrene/(sum of 1-,2-,3-,9-methylphenanthrene + sum of 1,3,5-, 1,3,6-, 1,3,7-, 1,4,6-, 2,3,6-trimethylnaphthalene) for coal samples from five different coal seams in the Ruhr area, western Germany (after Littke et al., 1990). Maturity is increasing from coal seam V to coal seam C (R_r = 0.70-0.92%). Arrows mark samples compared in Figure 65. Note the poor correlation between maceral group composition and chemical data measured on extractable bitumen at a given maturity level. This indicates a significant redistribution of these compounds in the open pore system of the coals (see Radke et al., 1990, for discussion of open and closed pores).

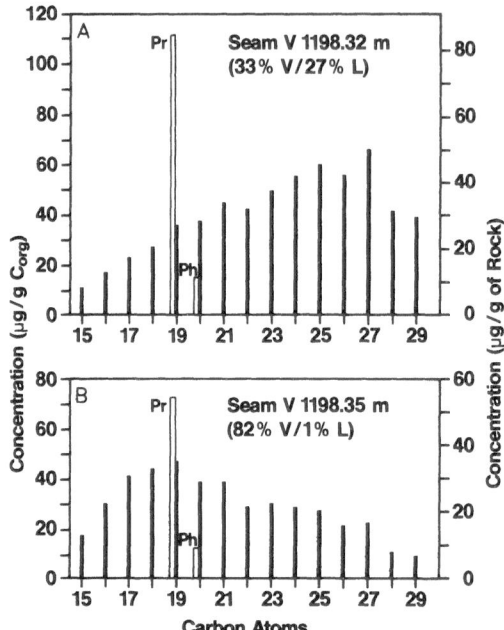

Fig. 65:   Concentration of C_{15+}-*n*-alkanes (solid bars) and pristane and phytane (open bars) in two selected vitrinite- and liptinite-rich coal samples from the V-seam (see Fig. 64). Regarding the difference in vertical distance between these samples (three centimeters), the difference in *n*-alkane distribution indicates a limited migration of these compounds at the relatively low maturity level reached by these samples (0.70% R_r).

Another key towards a quantification of migration efficiencies may be concentration data on pristane and phytane in coals of different maturities. In the sample series shown in Fig. 63, mean pristane/phytane concentration ratios decrease from 7.3 to 4.6 (Table 26). A previous study by Radke et al. (1980) also revealed a significant maturity-dependent decrease in pristane/phytane ratios for coals, but the onset was observed at 0.9% R_r instead of 0.7% R_r. In the Nesberg 1 well, the decrease of the pristane/phytane ratio is due to a depletion in pristane at rather constant phytane concentrations (Fig. 70; see Radke et al., 1980).

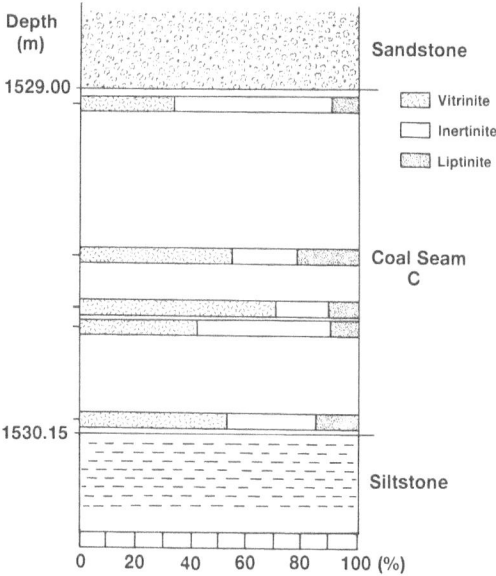

Fig. 66:   Maceral group composition of five samples from coal seam C (0.92% R_r, see Fig. 64)
              from the Ruhr area, western Germany (from Littke and Leythaeuser, in press). No
              obvious regular trend in maceral group composition was found for this seam, which is
              overlain by a sandstone (see Figs. 67-69).

Pyrolysis-GC measurements on samples from the uppermost interval of the sample series revealed that
prist-1-ene, a known pristane precursor (Goossens et al., 1988), is not present. Therefore, generation of
additional pristane does not seem probable in the studied rank interval so that the 50% decline of pristane
concentration (Fig. 70) is assumed to be only an effect of loss by migration. The constant phytane
concentrations have to be explained by a new generation of this compound that compensated for the loss
by migration. The alternative explanation, a preferential expulsion or cracking of pristane compared to
phytane does not seem reasonable in view of the similarity of both molecules and in view of the lack of
fractionation effects deduced for coal seam C. Thus, it is assumed that about 60 μg/g C_{org} pristane were
expelled during maturation from 0.70 to 0.92% R_r on an average. If additional pristane was generated
within this maturity interval, a greater amount of expelled pristane (and phytane) has to be assumed.

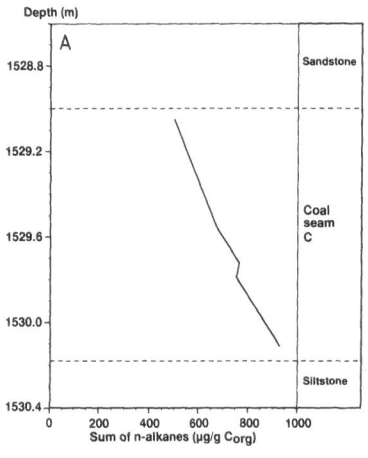
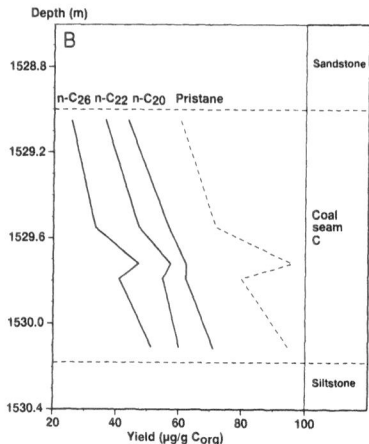

Fig. 67a,b: Yields of C_{15+}-*n*-alkanes (a) and individual *n*-alkanes and pristane (b) (µg/g C_{org}) in coal seam C. In view of the maceral group composition (Fig. 66), the regular decrease towards the sandstone is explained as related to different degrees of expulsion (after Littke et al., 1990).

If it is further assumed that the expulsion efficiency of pristane was the same as the expulsion efficiency of saturated hydrocarbons in general and that the original organic matter composition of seams C and V (see Fig. 64) were identical, the ratio of saturated hydrocarbons over pristane at the immature stage multiplied by 60 µg/g C_{org} provides the quantity of saturated hydrocarbons expelled from coal seam C during maturation from 0.70 to 0.92% R_r. The result of this calculation (see data in Littke et al., 1990) is that 1010 µg/g C_{org} or about 1mg/g C_{org} of saturated hydrocarbons were expelled. This number would only refer to the saturated hydrocarbons already present at the immature seam V stage. Expulsion of saturated hydrocarbons additionally generated in the rank interval between 0.70 and 0.92% R_r is not included in the above number.

In summary, geochemical data indicate that oil expulsion can occur from some coal seams at maturity levels of less than 1% R_r. Rough estimates on quantities of expelled products can be given only for

155

saturated hydrocarbons at the present time. For aromatic hydrocarbons, there are observations supporting an efficient redistribution within coal seams, but little evidence for expulsion is available.

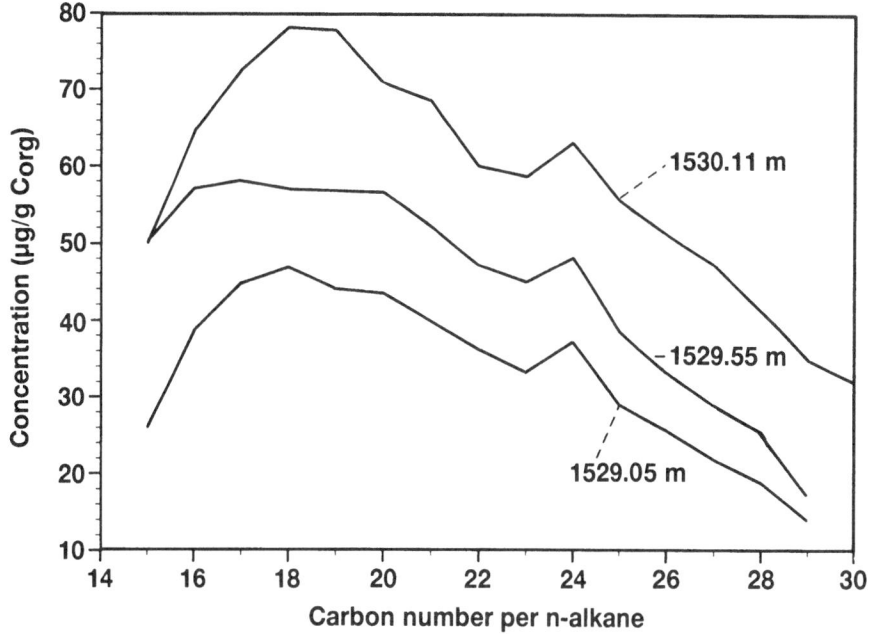

Fig. 68:   Concentration of C_{15+}-n-alkanes in three samples from coal seam C. The difference between the sample next to the sandstone and the samples in the central and lower part of the coal seam is tentatively interpreted as an expulsion phenomenon (after Littke et al., 1990).

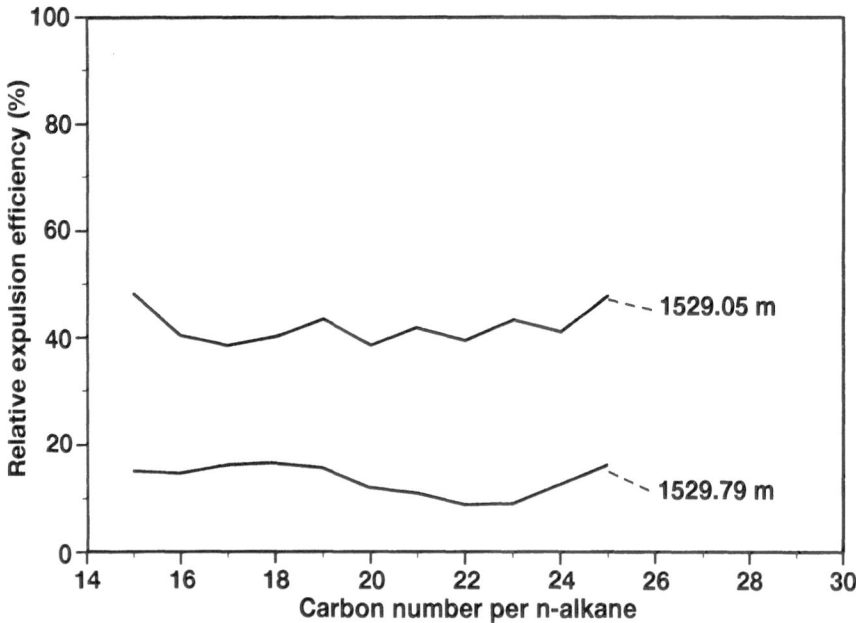

Fig. 69:   Calculated relative expulsion efficiencies (see Leythaeuser et al., 1984a for mode of calculation) for two different samples from coal seam C. The lowermost sample of this seam (1530.11m depth) was chosen as a reference sample (from Littke and Leythaeuser, in press).

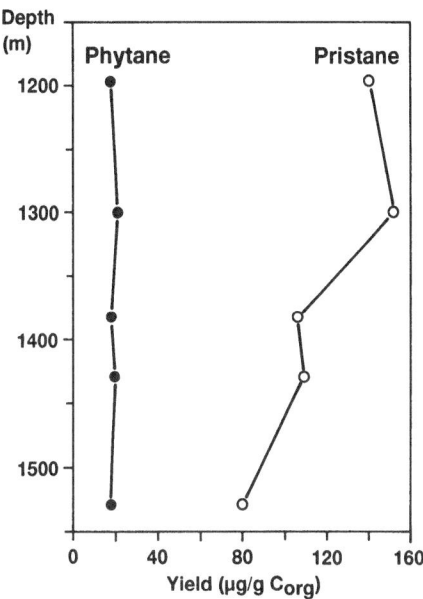

Fig. 70:   Average pristane and phytane concentrations of the coal samples ($\geq$ 60 % C_{org}) of
the five depth intervals shown in Fig. 63 (from Littke and Leythaeuser, in press).

7.5   Mass balance approaches

One major goal in any primary migration study is to determine the masses of volatile products expelled
from a source rock. As the volatile products are generally not available for measurement, these
quantifications have to be done on the basis of the elemental composition of the immobile residues, the
coals. Mass balances based on the evolution of the elemental composition of "normal" coals or specific
coal lithotypes were performed by several authors (e.g., Jüntgen and Karweil, 1966a; Boudou et al., 1984;
Littke et al., 1989). Prerequisites for the validity of these mass balances are that i) coals of similar original
(=pre-maturation) composition are compared and ii) no material was added to the coals.

Jüntgen and Karweil (1966a) described the evolution of coals from the bituminous coal to the anthracite
coalification stage ($R_r \sim$ 0.9 - 4.0%) using the assumption that CH_4, CO_2 and H_2O are the major expelled
volatile products. For the mass balance, they used an equation system (here modified):

$$C_i - \alpha C_r = 3/4x + 3/11y \qquad (1)$$

$$H_i - \alpha H_r = 1/4x + 1/9z \qquad (2)$$

$$O_i - \alpha O_r = 8/11y + 8/9z \qquad (3)$$

where $\underline{C}_i$, $\underline{H}_i$, and $\underline{O}_i$ are percentages of carbon, hydrogen and oxygen in the initial coal, $\underline{C}_r$, $\underline{H}_r$, and $\underline{O}_r$ are the percentages of the same elements in the residual coal, and $\underline{x}$, $\underline{y}$, and $\underline{z}$ are masses (g) of CH_4, CO_2 and H_2O expelled during maturation. The calculation is valid for 100 g of initial coal, which results in α ·100 g of residual coal.

Under the assumption that water will only be generated and expelled from coals at high maturity levels (> 3% R_r) when oxygen is only present in hydroxyl (OH-) groups, z in equations 2 and 3 is zero and x, y, and α, i.e., masses of expelled CH_4 and CO_2 can be calculated. In a second scenario, it is assumed that no carbon dioxide is expelled from coals with more than 92% C ($R_r > 3\%$). In this case, y in equations 1 and 3 is zero and x, z, and α; i.e., masses of expelled CH_4 and H_2O can be calculated. Results are presented in Fig. 71a and indicate that within the bituminous coal stage (0.9-2.0 % R_r), CO_2 expulsion exceeds CH_4 expulsion, whereas in the anthracite stage (> 2.0% R_r), methane expulsion is the most efficient process. Water is expected to be generated and expelled only at very high maturities (thin lines in Fig. 71a), but it should be noted that this is the result of an *a priori* assumption and not a result of these calculations.

In Figs. 71b and 71c, calculated masses of methane and carbon dioxide are plotted as a function of maturity for two maceral groups in coal, liptinite and vitrinite, respectively. Jüntgen and Karweil's (1966a) results indicate that methane generation from liptinite is far more important at low levels of maturation, whereas vitrinite behaves almost like an average coal (compare Figs. 71a and 71c). The latter finding is not surprising in view of the fact that an average coal consists of about 70% vitrinite (e.g., Littke, 1987). In the case of liptinite, it is reasonable to assume that considerable amounts of hydrocarbons other than methane are generated and that not only methane but also other hydrocarbons are expelled from the coals.

Jüntgen and Karweil (1966a) summarized their results integrating the calculated masses of CH_4 and CO_2, as well as of nitrogen (Fig. 72a). They suggest that carbon dioxide, which is soluble in water, is easily expelled from the coals and that accordingly the percentage of nitrogen is high in the coal gas at relatively low levels of maturity (Fig. 72b). This idea may bear some importance for the understanding of commercial natural gas fields, which are rich in nitrogen. Examples exist, for instance, in the southwest of the United States and in western Europe, where nitrogen percentages in gas reservoirs reach maximum values in excess of 80% (Tissot and Welte, 1984:211). Krooß et al. (in press) measured kinetic data on nitrogen and methane generation from coal and concluded that under specific heating rate conditions, significant nitrogen release from coals occurs within geological systems. These authors suggest that very high temperatures (>300°C) are necessary for preferential nitrogen generation in nature.

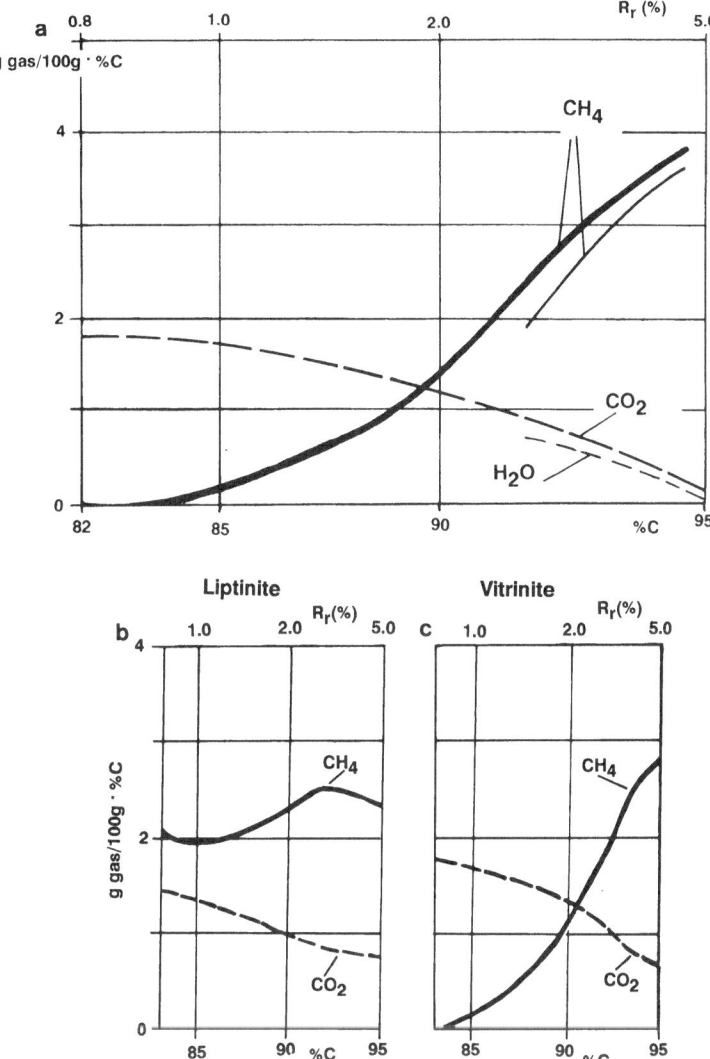

Fig. 71a,b,c: Calculated gas generation during natural coalification processes for coals in general (a), liptinite (b), and vitrinite (c) (from Littke and Leythaeuser, in press; after Jüntgen and Karweil, 1966a). Only liptinites generate more methane than carbon dioxide at low levels of maturation ($R_r < 2.0\%$).

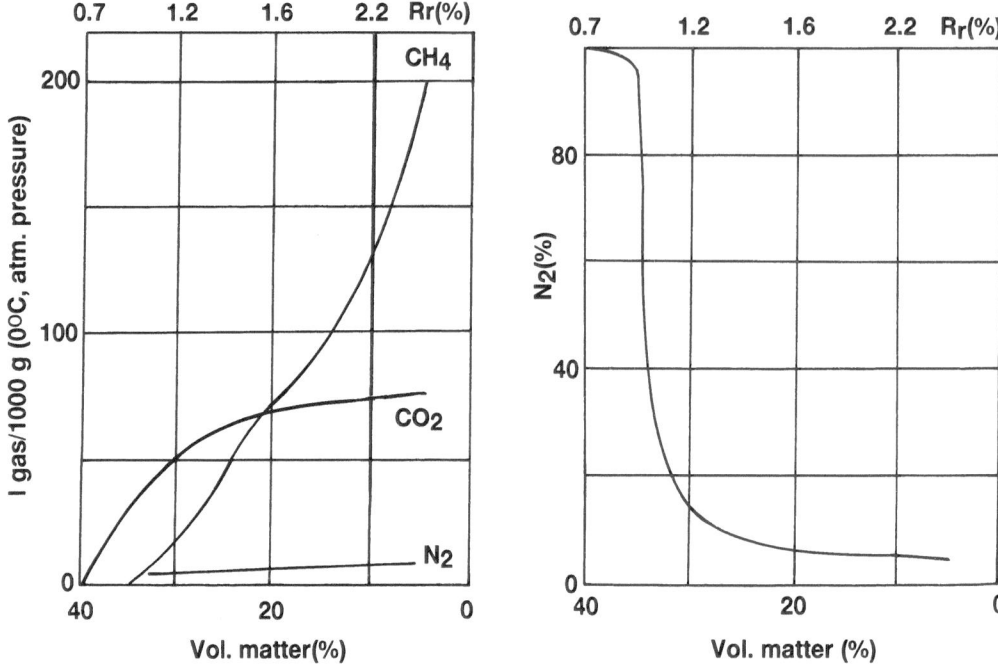

Fig. 72a,b: Calculated integrated gas expulsion (methane, carbon dioxide, nitrogen) in l/kg original coal (0.7% R_r coal) (a) and calculated nitrogen concentration in the coal gas (b) (from Littke and Leythaeuser, in press; after Jüntgen and Karweil, 1966b). It is assumed that carbon dioxide is easily expelled from the coal seams at low levels of maturation due to its great water solubility.

The total volumes of gases and liquids expelled form vitrinites and the mass loss of vitrinites are visualised in Fig. 73. For this diagram, elemental compositions of vitrinites were used (Stach et al., 1982: 89) Furthermore, the above mass balances (Boudou et al., 1984; Jüntgen and Karweil, 1966a; Littke et al., 1989) provided information on the probable gas composition generated and expelled at the different rank levels. The total mass loss of almost 50% implies that 50% of the nitrogen present in immature coals is expelled, if nitrogen percentages remain at the immature level (1.5-2.0% on average). As nitrogen percentages seem to decrease at very high maturities (metaanthracite stage, >95% C, >4% R_r), the nitrogen release may be even greater. Thus, about 1% of the immature coal mass can be expected to be lost as nitrogen gas, preferentially at high maturities and temperatures.

The elemental composition of less mature coals was used by Boudou et al. (1984) to calculate amounts of CO_2, H_2O and hydrocarbons expelled during diagenesis (< 80% carbon, < 0.7% R_r). They concluded that mainly carbon dioxide is lost at this stage, and that hydrocarbon production and expulsion is negligible.

Elemental compositions of liptinite-rich dull coals of similar organic facies were compared for the rank range between 0.70 and 0.92% R_r by Littke et al. (1989). This range coincides with the maximum oil generation stage as previously found by Radke et al. (1980, 1982). Methane generation is expected to be low at this maturity interval (Jüntgen and Karweil, 1966a; Radke et al., 1982). The approach to select only a few petrographically similar coals rather than "average" coal or whole coal samples was used in view of the great variety of chemical and petrographic composition of these coals. Weight percentages of carbon, hydrogen, and oxygen for two liptinite-rich coals are shown in Fig. 74a. Under the assumption that no water is expelled at that stage (Jüntgen and Karweil, 1966a; Battaerd and Evans, 1979), a loss of 14% of CO_2 and 8% of hydrocarbons (CH_2) was calculated, which results in a total mass loss of 22% within this small maturity interval (Fig. 74b). Nevertheless, only 20% of the hydrogen present at 0.70% R_r was lost during maturation to 0.92% R_r (see Littke et al., 1989 for more explanation).

Rullkötter et al. (1988) presented results for petroleum generation and expulsion from Lower Toarcian petroleum source rocks covering approximately the same maturity level. According to their mass balance, 27% of the original marine kerogen is already converted into volatile products (mainly hydrocarbons and carbon dioxide) and expelled at maturity levels below about 0.7% R_r. This compares to a significant carbon dioxide and water expulsion in coals, whereas no hydrocarbons are generated and expelled at this stage (Boudou et al., 1984).

In the maturity range between about 0.7 and 0.9% R_r, Rullkötter et al. (1988) found that additional 23% of the original marine kerogen or 32% of the kerogen present at 0.7% R_r are converted into volatile products and expelled. The expelled products are almost exclusively hydrocarbons in this interval. In comparison to coals in the same maturity interval (Fig. 74), the hydrogen and hydrocarbon loss is much greater in marine kerogen and the oxygen and carbon dioxide loss is much smaller (Fig. 75a and b). The hydrogen loss of the marine kerogen is, for example, 40% compared to only 20% for coals in the same maturity interval (0.70 - 0.92%).

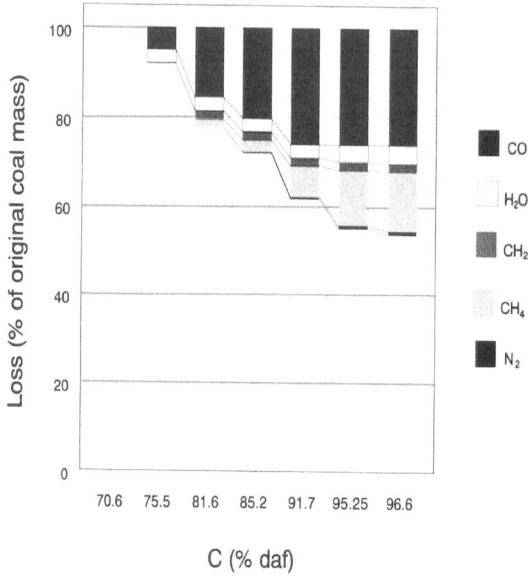

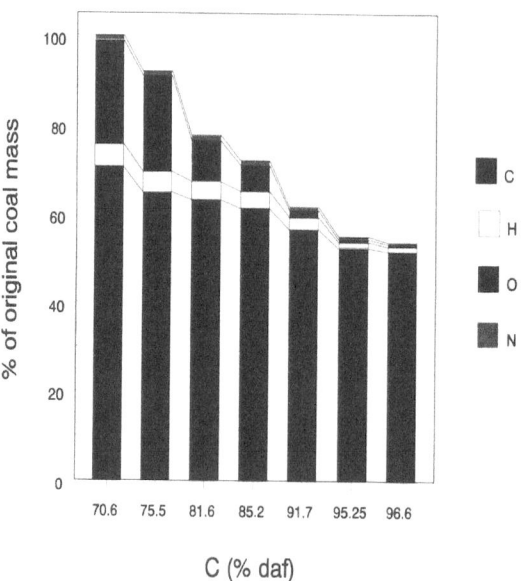

Fig. 73:   Evolution of the elemental composition of vitrinite and simulated evolution of cumulative gas composition and mass. Data on the elemental composition of vitrinite in Stach et al. (1982:89) were used for this simulation.

It should be noted that the mass balances for coals on one hand and for a marine source rock on the other are based on different procedures; therefore the comparison contains some pitfalls. First, it has to be assumed that vitrinite reflectance values measured on coals are equivalent to those measured on the Lower Toarcian source rock and adjacent marine strata. Secondly, the mass balance of Rullkötter assumes the same initial average organic carbon percentages for the Posidonia Shale in all wells which is supported by lithologic data. Nevertheless, the resultant mass balance reveals slightly different expulsion and generation data than the alternative approach by Cooles et al. (1986) which is purely based on laboratory experiments. Using the latter technique, the differences between coals and marine source rocks would be smaller than expressed by Fig. 75.

The above-cited mass balances certainly suffer from the fact that there is no exact knowledge of the products that are really expelled from coals. Therefore, all calculations are based on the simplification that only carbon dioxide, water and hydrocarbons (CH_2 or CH_4) are migrating within and out of the coals. In reality, a complex mixture of molecules will be expelled (see Chapter 7.4). However, the mass balance approach provides a unique tool to get quantitative information on the total mass loss that coal experiences and on the loss of individual elements. In this respect, coals obviously differ greatly from marine kerogen, because most hydrocarbons are only expelled at high levels of maturation ($> 0.9\%$ R_r). In marine source rocks, the bulk hydrocarbon generation and expulsion is already completed at that stage. In coals, more hydrogen is preserved until high maturity levels are reached. This is in general agreement with kinetic data which predict generation of hydrocarbons from coaly organic matter (type III kerogen) at higher temperatures than from marine source rocks with type II kerogen (Tissot and Espitalié, 1975).

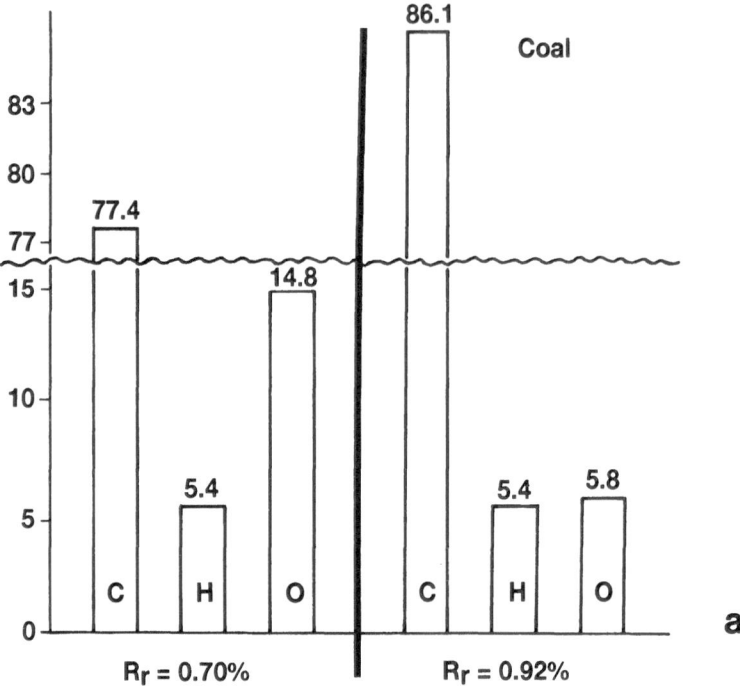

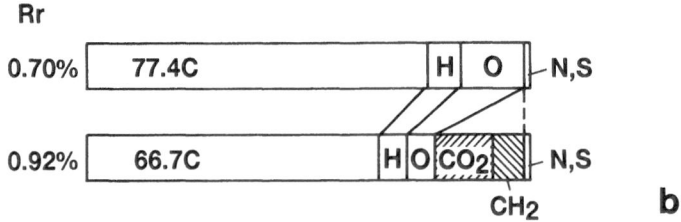

Fig. 74a,b: Comparison of the elemental composition of two sporinite-rich dull coals (same lithofacies, similar maceral composition) at different maturity stages (0.70% and 0.92%) (a). The difference can be expressed as a loss of 14% CO_2 and 8% CH_2 (hydrocarbons) (b). With the assumption of this loss of volatile compounds, the 0.70% coal would evolve towards an elemental composition similar to the 0.92% coal sample. Other (more complex) evolution pathways can also be assumed, but they all reveal a significant loss of volatile products (10 to 25%) and a loss of about 20% of the original hydrogen (after Littke et al., 1989 and Littke and ten Haven, 1989).

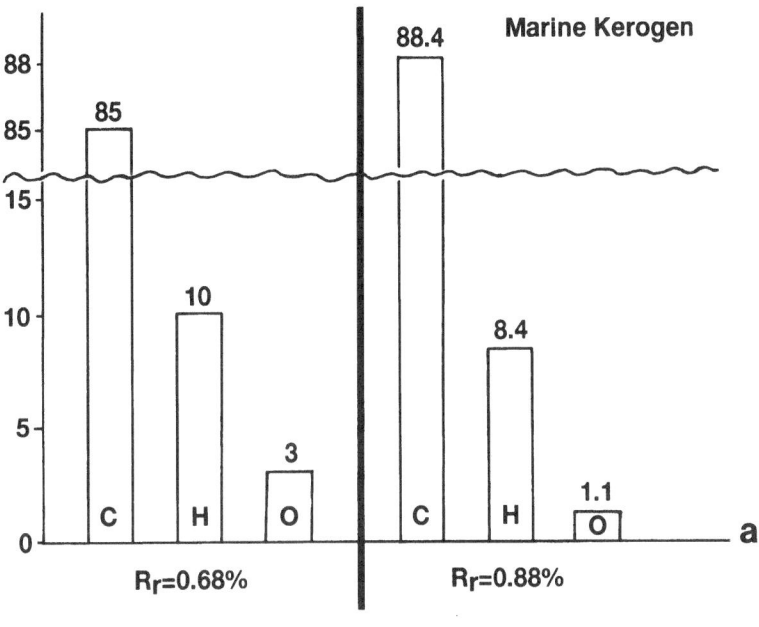

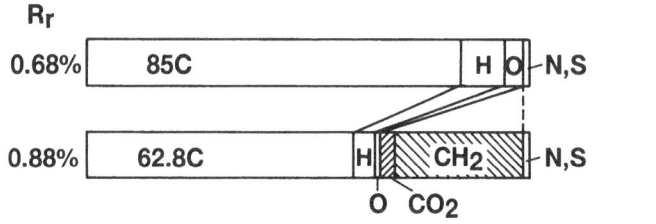

Fig. 75a,b: Comparison of the average elemental composition of marine kerogen in two wells (Dielmissen and Harderode, 0.68 and 0.88% R_r, respectively), (a) through the Lower Toarcian Posidonia Shale (from Littke and Leythaeuser, in press; based on data from Rullkötter et al., 1988a). The resultant loss of hydrogen (b) is about twice as high (40%) than in the case of coals in the same maturity interval. More hydrocarbons and less carbon dioxide are expelled (compare Fig. 74).

7.6   Migration of gas

For gas exploration, it is certainly of interest whether generated methane is trapped within the coal seams (coalbed methane) or expelled from those into adjacent rocks. Jüntgen and Klein (1975) calculated on the basis of experimental data that the total volume of methane that can be generated from a high volatile bituminous coal is in the range of 70-125 m^3 (at surface pressure and temperature) per ton. The corresponding volume of nitrogen is about 4 m^3 per ton (see Krooss et al., in press). The amount of gas that can be stored in coals by adsorption is usually calculated based on the specific surface area (Jüntgen and Karweil, 1966b).

Different theories (BET, Langmuir, Dubinin-Polanyi) exist to calculate specific surface areas, which are discussed and compared elsewhere (e.g., Debelak and Schrodt, 1979). Surface areas for high volatile bituminous coals are generally in the range of 150 to 350 m^2/g. Experiments reveal that methane adsorption in coals is positively correlated to pressure and maturity (Fig. 60; Janowsky, 1984) and negatively correlated with temperature. For example, at low levels of maturation (e.g., 0.7% R_r) and a constant pressure of 10 MPa, almost twice as much methane is adsorbed at low temperatures (29-50°C) than at high temperatures (100°C; Jüntgen and Karweil, 1966b). Thus highly mature coals, preferentially anthracites, situated at great depth in areas of low recent heat flow should contain the greatest amounts of adsorbed gas. It is, however, believed that additional factors such as organic facies (maceral composition) or tectonic stress will also influence methane adsorption in coals.

Gas is not only adsorbed on coal surfaces, but occurs also in the pore volume of coals. Its amount may exceed the amount of adsorbed methane, if methane is assumed to fill the entire coal porosity. It is, however, not well known to what extent large pores measured under laboratory conditions exist in the subsurface (see section 7.2). No matter whether it is assumed that pore filling methane significantly adds to the adsorbed methane or not, the total storage capacity of coals is exceeded by methane generation already at levels of maturation corresponding to a maturity of about 1% vitrinite reflectance (Jüntgen and Klein, 1975).

An attempt to quantify methane expulsion from coals in comparison to that from terrigenous organic matter dispersed in interbedded shales and mudstones was published by Scheidt and Littke (1989). The authors first determined the percentages of coal, mudstone, siltstone, and sandstone in the upper part of the coal-bearing Carboniferous strata of the Ruhr area (Fig. 76a). In addition, the average organic matter content of these four lithologies was quantified and vitrinite reflectance was measured for maturity assessment. Subsequently, methane generation and expulsion were calculated based on the experimental data and theoretical concept of Jüntgen and Klein (1975) on methane generation and storage capacities. No storage capacity was assumed for dispersed kerogen in clastic rocks.

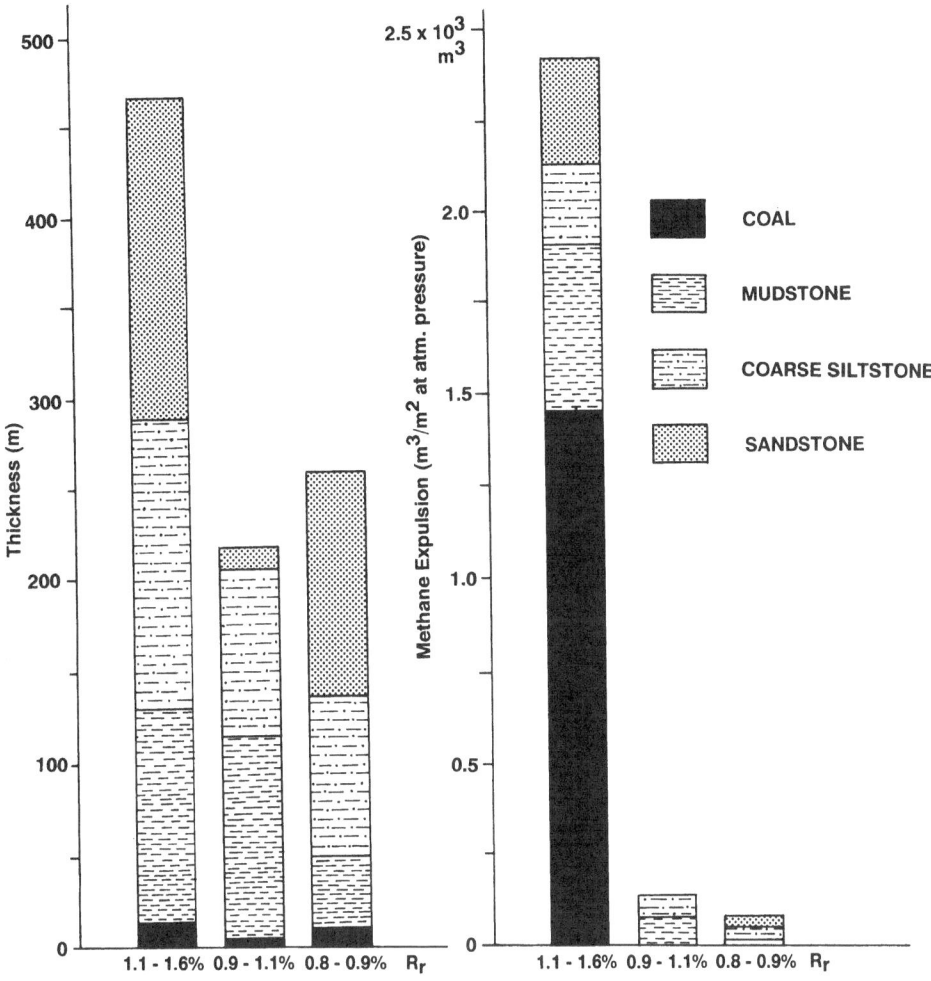

Fig. 76a,b: Thickness of four basic lithologic units (coal, mudstone, siltstone, sandstone) in three stratigraphic and rank intervals of the Ruhr basin, western Germany (a) and calculated methane expulsion from these same lithofacies (b). According to the formalisms used, methane expulsion from coals is efficient only at maturities corresponding to $R_r > 1.1\%$ (after Scheidt and Littke, 1989).

The resultant expelled masses of methane are shown in Fig. 76b. At low levels of maturation ($R_r \sim 0.8$ - 0.9%), no methane expulsion from coals and only a small expulsion from dispersed kerogen was found. In reality, volatile products, including hydrocarbons, will be expelled from the edges of coals at this stage, but in small quantities. There is evidence to suggest that hydrocarbons other than methane are generated and partly expelled at this low maturity (see last section). At higher levels of maturation ($R_r \sim 0.9 - 1.1\%$), only slightly greater methane expulsion is calculated for dispersed kerogen and no expulsion for coals, because storage capacities are not yet exceeded by methane generation. This is, however, not the case for the deepest stratigraphic interval ($R_r \approx 1.1 - 1.6\%$, see Scheidt and Littke, 1989), in which great amounts of methane are expelled from coals. The expulsion from dispersed kerogen is lower, because more organic matter is stored in coal seams than in clastic rocks in the Ruhr area. In the above calculations, the potential for methane generation was considered to be equal for coals on one hand and dispersed terrigenous kerogen on the other. This is suggested by microscopic countings, but not supported by chemical analyses. The latter reveal a significantly lower hydrocarbon generation potential for dispersed kerogen than for coals of the same maturity as reported by Littke and ten Haven (1989) for the Ruhr area and by Huc et al. (1986) and Horsfield et al. (1988) for other coal-bearing basins. Thus, the methane expulsion from dispersed kerogen may have been overestimated in Fig. 76b.

In summary, gas is expelled from coals at maturity levels exceeding about 1.0% vitrinite reflectance at about the same rate as it is generated. In most basins, the methane generation and expulsion capacity of a given mass of coal exceeds that of the same mass of dispersed terrigenous organic matter (type III kerogen).

7.7   Conclusions

In the previous chapters evidence was summarized to suggest that liquid hydrocarbons are generated in coals and partly expelled. The bulk of the original hydrogen does, however, remain in the coals until high maturity levels are reached, upon which conversion into gas, mainly into methane, is to be expected. All methane (and hydrogen) is expelled from the coals during maturation from the bituminous coal to anthracite, meta-anthracite and finally graphite stage. Thus, coals can be regarded as moderate to poor quality oil source rocks. For practical applications in petroleum exploration, the individual petrographic and geochemical characteristics of coals have to be studied in order to judge their oil generation potential. Liptinite-rich coals which produce long-chain n-alkanes upon pyrolysis are expected to possess a high oil generation potential. In addition, the permeability of coals can be used to get insight into the expulsion efficiency. Uplifted or fractured coals are regarded as better prospects than deep, unfractured coals in this context. Coals are, excellent gas source rocks, because coal seams are extremely rich in organic matter and keep much of the initial hydrogen until high maturity levels and optimum conditions for generation of gas are reached.

The similarity of compositions of C_{15+}-saturated hydrocarbons and C_{11+}-aromatic hydrocarbons between different coal lithotypes <u>within</u> one coal seam suggests an effective migration of oil within the coal seams. This contrasts with the mass balance results based on elemental composition, which indicate a limited expulsion of hydrogen-bearing compounds at low levels of maturation ($< 1.0\%$ R_r). Limited expulsion of liquid hydrocarbons can be interpreted as an effect of the coal chemical structure, e.g., predominance of methyl groups over n-alkyl groups, which would lead to a limited generation of oil (see Fig. 57). Another possible explanation is the existence of "open" and "closed" pores (Beletskaya et al., 1976; Radke et al., 1982, 1990). The "open" pores (macropores, mesopores; see section 7.2) allow expulsion and are accessible for non-specific solvents, whereas the "closed" pores do not permit migration and are non-accessible to solvents. "Closed" pores (micropores) are expected to be most abundant in vitrinite (Harris and Yust, 1976); therefore, the expulsion efficiency for oil is assumed to be higher for vitrinite-poor bituminous coals than for the more typical vitrinite-rich bituminous coals.

8. WEATHERING OF POSIDONIA SHALE IN NORTHERN GERMANY

8.1 Overview

Weathering is defined as the sum of processes that change the organic and inorganic constituents of rocks in contact with atmosphere, hydrosphere, and biosphere (Valeton, 1988). Generally, these processes also change the pore systems of rocks. Weathering is known to affect amount and quality of organic matter in petroleum source rocks (Baker, 1962; Leythaeuser, 1973; Clayton and Swetland, 1978; Forsberg and Bjorøy, 1981; Clayton and King, 1987). These changes may influence petroleum generation capacities as well as the maturity indicators measured by means of organic geochemistry and organic petrology.

In the study of Littke et al. (1991d), it was the primary objective to quantitatively evaluate the weathering-related mass loss of organic matter and to give more insight into qualitative aspects of the degradation of organic matter. This was achieved by comparing two series of Posidonia Shale (Lower Toarcian) samples from the Hils area, Northern Germany. One of the series was cored below 29 m of overlying sediments (borehole Wenzen), the other series is derived from three different outcrops about 150 m south of the Wenzen well (Fig. 77). It is reasonable to assume that the original (= pre-weathering) average rock composition and especially the organic matter-yield, -composition, and -maturity is identical for both sample series.

The second objective was to study how weathering affected minerals in the Posidonia Shale and to calculate a mass and volume balance describing Posidonia Shale weathering. Therefore, data on average percentages of silicates, carbonates, and sulphides had to be incorporated into the calculations and are documented in this text. In addition, a few pore volume and pore size distribution data are used to test the conclusions.

Rocks in the Hils area are generally weathered to a depth of about 5 m as evident from unpublished data on shallow cores. For calculations of bitumen and sulphur loss rates, two different weathering ages we used. The weathering ages here are defined as the times at which samples in the centres of the outcrop were sufficiently exposed to the surface that weathering started. The resultant weathering rates are designated to approximate natural release of bitumen and sulphur from the shale into the hydrosphere which can then be compared to rates of anthropogenic processes.

Seventy Posidonia Shale samples were collected at three separate outcrops in the southern part of the Hils area in northern Germany. Thirty five of these samples were taken from the surface (outcrop face) and the other thirty five at the same locations after digging about 30 cm into the rock. As the geochemical data revealed, no significant differences exist between these two sets of outcrop samples. Therefore, both are treated together and referred to as outcrop samples or weathered samples in this text. Within the three outcrops, no systematic changes in chemical properties of the rocks from top to bottom were found. However, the Posidonia Shale from outcrop 1 which is an old erosion depression appeared macroscopically less severely weathered than the Posidonia Shale from outcrops 2 and 3 which are located in a railway cut constructed in 1865. Therefore, average data of some organic geochemical parameters were calculated separately for outcrop 1 and for outcrops 2 and 3.

The outcropping Posidonia Shale was compared to Posidonia Shale from the shallow, completely cored borehole Wenzen 1001 situated about 150 m north. Here, the Toarcian black shale is overlain by 24m of Dogger and 5m of Quaternary deposits, i.e., it has never been exposed to the surface and has not been affected by weathering. The Posidonia Shale core consists of a geochemically homogeneous upper calcareous shale (11 m) and a lower marlstone (4m) unit (Fig. 77, Littke and Rullkötter, 1987). On the basis of lithologic criteria, the stratigraphic interval which is exposed in all three outcrops is the upper (less calcareous) shale facies. From this interval of the core 45 samples were collected. The same analyses were completed on both outcrop and core samples, and are described by Rullkötter et al. (1988a) and Littke et al. (1988).

8.2   Mass and volume balance: Definitions and assumptions

The quantification of the weathering process during the final phase of uplift, preceding erosion must be based on comparison of "unweathered" and "weathered" samples of the same facies and maturity. Initially it is assumed that weathering is not associated with compaction and that, consequently, a unit volume of the weathered rock corresponds to a unit volume of the unweathered material. This assumption is supported by petrological evidence, in particular the intact petrological appearance of outcrop samples regarding the similar average distance between common petrographic constitutents such as fluorescing alginite in outcrop and core samples of the Posidonia Shale.

In order to quantify the effects of weathering, five different fractions are separately described: 1) silicates, 2) carbonates, 3) sulphides 4) organic matter, and 5) iron oxide, an important weathering product in the outcrop samples. These five fractions represent almost 100% of the petrographic inventory of the Posidonia Shale.

Silicates

In the core samples, silicates consist of quartz and clay minerals, among which illite, kaolinite, and minor illite/smectite mixed layers prevail (Mann, 1987). XRD-traces of the < 2 μm fraction of weathered outcrop samples revealed that the clay mineralogy is similar to the clay mineralogy of the cored Posidonia Shale. The only difference between outcrop and core samples is a higher relative portion of unordered smectite/illite mixed layer minerals in outcrop samples. The < 2 μm fraction is thought to be more labile than the > 2 μm fraction and therefore best suitable for identification of mineralogic changes. Due to the small compositional difference between the clay mineralogy of < 2 μm fractions of outcrop and core samples, it is in a first scenario assumed that the silicate fraction was not affected by weathering and that this petrographic component can be used as a reference for comparing the weathered and unweathered state of the Posidonia Shale.

In order to investigate the consequences of a minor loss of silicate during weathering, a second scenario was considered, assuming a loss of 5% of the silicate originally present in a reference volume.

The relative silicate content ("reference fraction") of the samples was determined from the difference between total dry rock mass and the masses of the other petrographic components (Table 27). For the calculations an average density of 2.2 g/cm^3 was assumed for the silicate component.

Carbonates

Calcite is the only major carbonate mineral in both the core and outcrop samples. Petrographically, calcite mainly consists of coccoliths and other plankton-derived microfossils. A subordinate fraction of the carbonate is recrystallized. In the northern Hils syncline, however, where thermal maturity is much greater (Rullkötter et al., 1988a), recrystallized carbonate predominates over microfossil remains. The average percentage of carbonate is 37% in the core and 32% in the outcrops (Table 27). The difference is attributed to the loss of part of the originally present calcite. The general process of carbonate dissolution in freshwater can be expressed by $CO_2 + H_2O + CaCO_3 \rightarrow Ca^{++} + 2\,HCO_3^-$.

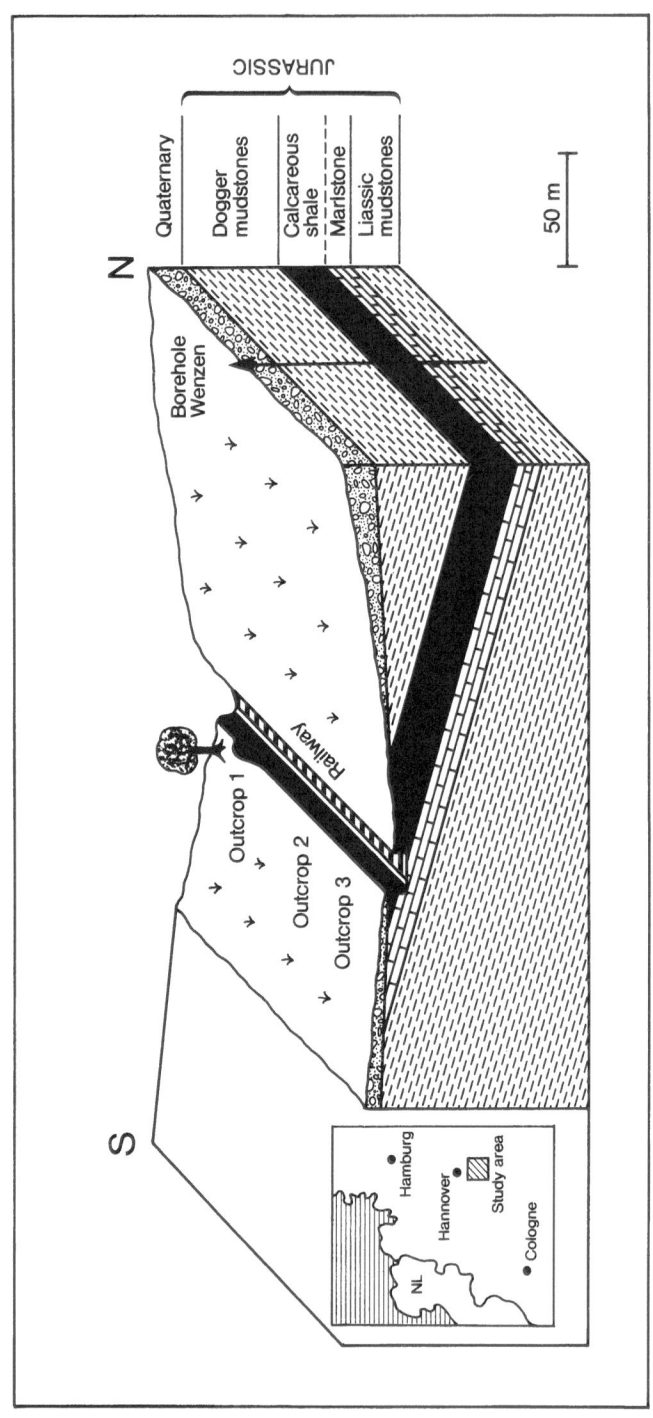

Fig. 77:   Schematic illustration of the location of borehole Wenzen 1001 and outcrops of
Posidonia Shale in the Hils area, northern Germany (from Littke et al., 1991d).

CO_2 release from organic matter due to remineralization in the black shale ($2\ CH_2O \rightarrow CO_2 + CH_4$) may be envisaged as an auxiliary process enhancing carbonate leaching.

The carbonate content of the shale samples was determined from the difference between total carbon (TC) and total organic carbon (TOC, C_{org}) assuming calcite ($CaCO_3$) with a density of 2.8 g/cm^3 as the only carbonate mineral present.

Sulphides

Pyrite (FeS_2) is the major sulphur-containing compound in this sedimentary sequence.

In the core samples, the bulk of the pyrite occurs as small ($< 5\ \mu m$) crystals which are either isolated or - more commonly - form aggregates of 10 - 20 μm diameter (framboidal pyrite). Comparison of the sulphur and iron contents of these samples (Table 27) reveals a slight excess of sulphur with respect to iron if the iron is assumed to be present exclusively as pyrite. It is assumed that this excess sulphur resides in the organic matter and that the percentage of sulphur in the organic matter does not change during weathering.

In the outcropping Posidonia Shale, much less pyrite is present. Many pyrite crystals seem to be completely replaced by iron oxide or iron hydroxide, but a small percentage appears unchanged. Occasionally, former pyrite crystals contain pyrite only in the centre which is surrounded by a dull, red-brown rim consisting of iron oxide or iron hydroxide. The microscopic appearance of pyrite seems to be a valuable and perhaps the most reliable indicator of source rock weathering, because pyrite is a ubiquitous constituent in most organic matter-rich sediments. The occurence of iron hydroxide replacing iron sulphide certainly does not indicate whether and to what extent organic matter is changed by weathering. It is expected, however, that organic matter in source rocks which contain exclusively fresh pyrite is not affected by weathering.

Organic matter

In the core samples most of the organic matter occurs as hydrogen-rich alginite and bituminite derived from marine planktonic organisms as described by Teichmüller and Ottenjann (1977) for the Posidonia Shale from other locations and quantified by Littke et al. (1988) for the Hils syncline area. A similar petrographic composition was established for the outcrop samples (Table 28). The average quantity of organic matter is calculated as 14.4% both in the core and in the outcrop series (Table 29). For the volume balance calculations a density of 1.2 g/cm^3 was assumed for the organic matter.

	Core	All Outcrops	Outcrop 1	Outcrops 2+3
$CaCO_3$ (%)	37.2±4.4 (45)	32.4±4.5 (70)	32.3±4.0 (28)	32.5±4.9 (42)
C_{org} (%)	10.6±1.6 (45)	10.6±1.5 (70)	10.2±1.2 (28)	10.9±0.8 (42)
S (%)	3.5±1.1 (20)	1.1±0.4 (70)	1.1±0.6 (28)	1.2±0.2 (42)
Fe (%)	2.6±0.2 (10)	3.0±0.4 (10)	3.1±0.3 (5)	3.0±0.5 (5)

Table 27: Geochemical data of the less calcareous upper facies of Posidonia Shale in the Wenzen core (depth range 29.9 - 40.4 m) and in three outcrops (from Littke et al., 1991d). Average weight percentages of calcite ($CaCO_3$), organic carbon (C_{org}), sulphur (S), and iron (Fe) are given with standard deviations and number of analyses (in parentheses).

Speciation of iron

Iron, one of the major elements in the mineral composition, was determined separately and also integrated into the mass balance. As evident from Table 27, Fe constitutes 2.63 wt.% of the dry rock in the unweathered core whereas its relative amount is slightly higher in the outcrop samples (3.03 wt.%; Table 27). A speciation based on the sulphur and iron contents of the samples and the assumption that iron is exclusively occurring as pyrite, indicates an excess of sulphur with respect to iron in the unweathered samples (see above).

For the outcrop samples which, in addition, contain significant amounts of iron oxide (presumably FeOOH), the speciation was made with the following assumptions:

- the percentage of sulphur residing in the organic matter ($m_S(OM)/m_{OM}$) is the same as in the unweathered core samples

- the sulphur which is not attributed to the OM occurs as pyrite (FeS_2)

- the remaining iron (not present as pyrite) is present as FeOOH with a density of 4.3 g/cm^3

Details on the calculation of the amounts of iron compounds in the shale samples are given in Appendix I.

Location	Vitr.+Inert. (Vol.-% of OM)	Bituminite (Vol.-% of OM)	Alginite A (Vol.-% of OM)	Alginite B (Vol.-% of OM)
Outcrop 1 (28)	5.8±4.6	9.3±4.0	13.8±3.8	71.1±7.0
Outcrops 2+3 (42)	3.5±2.4	10.1±4.1	11.4±5.1	74.1±6.9
Wenzen-core (12)	4.1±2.4	3.4±2.3	9.3±3.0	83.2±5.1
South-German Cores[1] (44)	6.6±14.3	5.6±5.4	6.7±5.8	81.1±14.7

Table 28:   Average volume percentages and standard deviations of terrestrial-derived macerals vitrinite and inertinite (Vitr. + Inert.), bituminite, large (> 20 μm) fluorescing particles (Alginite A), and small (< 20 μm) fluorescing particles (Alginite B) in Posidonia Shale (less calcareous upper facies) expressed as volume percentages of total macerals (from Littke et al., 1991d). Numbers of samples studied are in parentheses. Differences between average bituminite contents of the Wenzen core and the outcrops are probably due to different individual judgements of macerals during counting rather than to new bituminite generation during weathering. [1] Data from Rotzal (1990) for the entire Posidonia Shale.

8.3   Mass and volume balance: Results and discussion

The results of the mass and volume balance calculations are summarized in Table 29 and Figures 78A and 78B. By definition, a constant silicate mass is assumed in scenario I for a unit volume of Posidonia Shale and a loss of 5% of the original silicate is assumed in scenario II. For these scenarios, a loss of 23 wt.% (I) and 26 wt.% (II) of the original carbonate is calculated. For organic matter, relative losses of 11 wt.% and 16 wt.%, respectively, are calculated. Literature data by Leythaeuser (1973) and Clayton and Swetland (1978) revealed much higher losses of up to 25% and 60% for Upper Cretaceous and Permian shales from Utah, respectively. On the other hand, Clayton and Swetland (1978) and Forsberg and Bjoréy (1983) found no loss of organic carbon due to weathering in samples from the Cretaceous Pierre Shale in Colorado and samples from the Triassic Botneheia formation in Svalbard, respectively. Although the above cited previous studies dealt with smaller sample sets and did not incorporate the mineral matter loss into their calculations of organic matter loss, it is tentatively concluded that extreme weathering (25% and more loss of organic matter) is restricted to hot climates. It is, however, obvious from the difference between the two

Utah shales and the Colorado shale that additional factors such as permeability and reactivity of organic matter influence the weathering-related organic matter loss.

The average pyrite content was calculated as 5.64 wt.% for the core and 1.15 wt.% for the outcrops (Table 29). The difference is explained by a relative loss of 82 and 83 wt.% of pyrite for scenarios I and II, respectively (Fig. 78). Thus, pyrite is the primary petrographic component in the Posidonia Shale which is most strongly altered by weathering processes. One kilogram of this rock has lost 24 g of (pyritic) sulphur on average. Most of the sulphur was removed from the Posidonia Shale, probably as sulphuric acid (hydrogen sulphate) in aqueous solution. A small proportion remained as pyrite in the rock. Also, some gypsum visible as white "skin" on parts of the outcrops was precipitated. The low amount of gypsum is certainly an effect of the humid climate; weathering in arid areas would result in a much greater amount of gypsum.

In contrast to sulphur content, iron percentages are higher in the outcrop samples than in the core samples (Table 27). This may be explained by fixation of iron released from pyrite in iron hydroxides and an additional supply of iron by ground-water that was also fixed as iron hydroxide. The mineralogical and chemical composition of this weathering product is not known, but a goethite ($FeOOH$) composition is suggested. High organic carbon contents in cool, humid climates are known to favour the formation of goethite (Valeton, 1988). The calculated weight and volume percentages of goethite are summarized in Table 29.

Pore volume measurements were used to test the calculated mass loss for scenarios I and II. Data on pore volume and pore size distribution of Posidonia Shale from cores and analytical procedures are discussed by Mann (1987). Unfortunately, only few Hg-porosimetry measurements could be completed (Table 30) due to the difficulty in obtaining good quality drill plugs out of the shale. According to the average values of these measurements, an increase in pore volume from 14.6% in the cores to 26.6% in the outcrops was established (Table 30). The mass loss of calcite, pyrite, and organic matter calculated in scenario I results in a predicted final porosity of 23% whereas scenario II assuming 5% silicate loss during weathering yields a final porosity of 26.8% (Table 29). Considering the uncertainties involved with the analytical data used in these mass and volume balance calculations and the porosity measurements the calculated final porosities show a satisfactory agreement with the measured values and lend support to the assumption that not more than 5% of silicate was lost during weathering.

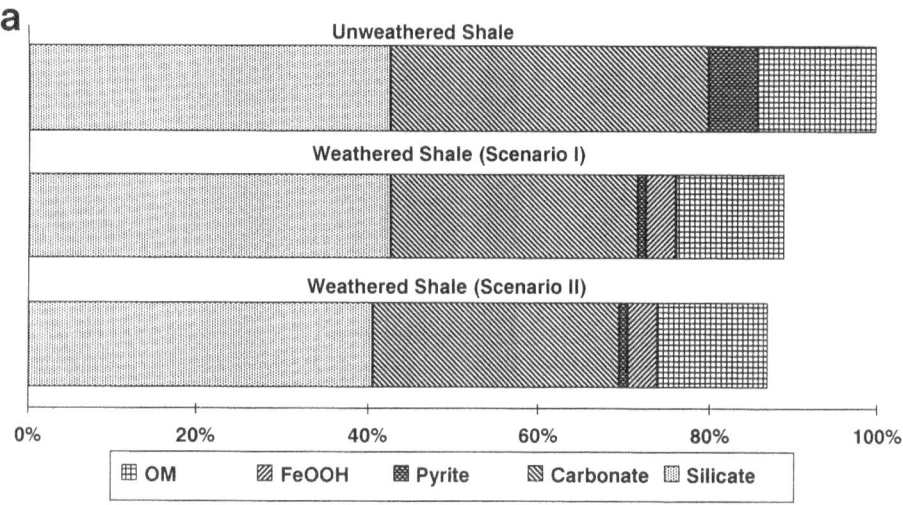

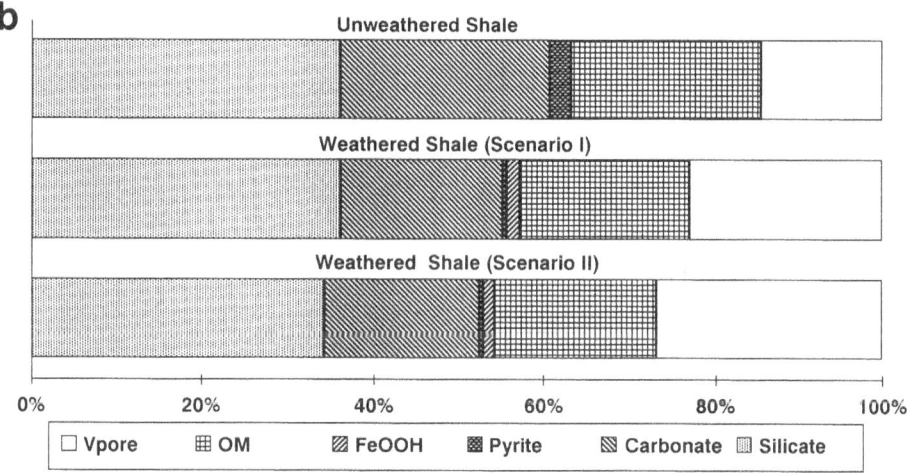

Fig. 78:   Mass balance (a) and volume balance (b) for scenarios I and II of weathering process of outcropping Posidonia Shale (from Littke et al., 1991d).

Petrographic component	Density g/ccm	Init.Content wt %	Weath. Content wt %	Rel. Loss (I) wt %	Rel. Loss (II) wt %	Init. Vol. vol %	Weath. Vol. (I) vol %	Weath. Vol. (II) vol%
Carbonate	2.8	37.20	32.40	23	26	25	19	18
Pyrite	4.5	5.64	1.15	82	83	2	0	0
Goethite	4.3	0.00	3.97	0	0	0	2	1
Org. Matter	1.2	14.40	14.40	11	16	22	20	19
Silicate	2.2	42.76	48.08	0	5	36	36	34

Table 29:   Calculated masses, volumes, and mass losses for different components in the less calcareous upper facies of the Posidonia Shale (from Littke et al., 1991d). Numbers (I, II) refer to the two scenarios discussed in the text.

	Core	Outcrop
Pore Volume (μl/g rock)	8.1 (2)	12.1 (4)
Porosity (%)	14.6 (2)	26.6 (4)
Largest Pores (nm)	118.0 (2)	222.0 (4)

Table 30:   Pore volume and porosity of core and outcrop samples of the less calcareous upper facies of the Posidonia Shale (from Littke et al., 1991d). Number of analyses in parentheses.

8.4   Organic matter characteristics

Hydrogen Index (HI) values from Rock-Eval pyrolysis average 612 mg hydrocarbons/g C_{org} for the outcrop samples and 666 mg hc/g C_{org} for the core (Table 31, see Rullkötter et al., 1988a). In the macroscopically less weathered outcrop 1, the average value is only 558 mg hc/g C_{org}, whereas it is 648 mg hc/g C_{org} in outcrops 2 and 3. Also, C_{org} values are lower in outcrop 1 (10.2) than in outcrops 2 and 3 (10.9). Thus, the macroscopic appearance is not a good indicator for estimating the degree of weathering in outcrop samples.

Figure 79 shows the relationship between organic carbon and pyrolysis yields (S_2) for the Wenzen core and for the outcrops. The similarity between the regression lines for outcrops 2 and 3 and for the core indicates that during weathering the quality of the organic matter was not significantly changed at these locations, although there was a loss of some organic matter as discussed in the previous chapter. Samples from outcrop 1, however, release lower amounts of pyrolysis products for a given organic carbon content than samples from the core suggesting lower H/C ratios in the organic matter (Espitalié et al., 1977). The different regression line for the samples from outcrop 1 is tentatively interpreted as an indication for significant changes of composition and structure of the organic matter as a result of either oxidation or preferential hydrogen release during weathering, i.e., the sum of the products released probably had a higher H/C ratio than the average of the original kerogen. Thus, the petroleum generation potential expressed as HI-value has been reduced by about 16% $((1 - 558{:}666) \cdot 100)$ at outcrop 1. The macroscopically most intact (unweathered) appearance contrasts with the most severely changed geochemical composition of the organic matter at this locality.

The general petrographic character of the organic particles is similar for core and outcrop samples (Table 28). Predominant macerals are fluorescing particles smaller than 20 μm in diameter (alginite B) which are thought to be derived from the nannoplankton of the Toarcian sea. Less abundant are larger fluorescing particles (< 200 μm, alginite A), low reflecting, weakly fluorescing particles (bituminite, < 500 μm) and brightly to moderately reflecting particles of terrigeneous origin (inertinite and vitrinite; usually less than 20 μm). It should be noted that in the case of Posidonia Shale about 20 volume-% of the rock consists of microscopically visible organic particles (= macerals; Littke and Rullkötter, 1987). As this number corresponds to 14.4 wt.% organic matter (Table 29) it can be concluded that the bulk of the organic matter is present as microscopically visible particles, i.e., there is only little submicroscopic organic matter (Fig. 3).

The most significant difference between macerals in outcrop and core samples is the darker fluorescence colour of many but not all alginites in outcrop samples. Often, areas with alginites showing dark fluorescence at longer average wavelength and areas with alginites of bright fluorescence at shorter average wavelength are seen within a single polished section of Posidonia Shale outcrop samples. In the core, only alginite of bright fluorescence at short wavelengths (λ_{max} = 530) was observed (Littke and Rullkötter, 1987). λ_{max}-values (wavelength of maximum fluorescence intensity) and Q-ratios (ratio of fluorescence intensities at 650 and 500 nm) of weathered samples are significantly different from those of unweathered cores (Fig. 80A+B). This change is of relevance in maturity studies, because fluorescence spectra parameters are often used as rank indicators (e.g., Fig. 37; Teichmüller and Ottenjann, 1977, Stach et al., 1982). In an earlier study on Posidonia Shale from core samples from a more mature well, it was shown that the considerable scatter of fluorescence parameters was possibly due to the different degree of impregnation of alginite with bitumen (Fig. 53; Littke et al., 1988). Correspondingly, the observed changes in fluorescence properties of alginites in outcrop samples are tentatively explained by loss of bitumen that is attached to alginites in core samples and was partly lost from alginites in outcrops due to weathering.

	Core	Outcrop 1	Outcrops 2+3
Hydrogen Index (mg/g C_{org})	666±34 (45)	558±80 (28)	648±42 (42)
S_1 (mg/g rock)	0.92±0.56 (45)	0.81±0.29 (28)	1.01±0.21 (42)
T_{max} (°C)	424±3 (45)	423±3 (28)	425±3 (42)
Soluble Organic Matter (mg/g C_{org})	39.2*±4.7 (10)	29.1±4.3 (9)	36.7±6.6 (10)
Saturated Hydrocarbons (mg/g C_{org})	6.3*±0.6 (10)	4.8±0.4 (9)	5.5±0.6 (10)
Aromatic Hydrocarbons (mg/g C_{org})	9.6*±1.0 (10)	4.0±0.9 (9)	5.9±1.5 (10)

Table 31:   Organic geochemical data (average values ± standard deviations) of the less calcareous upper facies of Posidonia Shale in the Wenzen core (depth range 29.9 - 40.4 m) and in three outcrops (from Littke et al., 1991d). Number of analyses are in parentheses.
* From Rullkötter et al. (1988a).

On average, the total amount of soluble organic matter in the core samples is 16% higher than in the outcrop samples (Table 31). Specifically, in the weathered samples the yield of saturated and aromatic hydrocarbons is reduced by 18% and 48%, respectively. Also, the molecular compositions of soluble organic matter differ considerably between outcrop and core samples. For example, the average yields of pristane and phytane are both about 20% lower in the outcrop samples, whereas the average yield of C_{15+}-n-alkanes is only about 10% lower. Within the latter group of compounds, a difference between core and outcrop samples can be clearly established for the pentadecane - heptadecane range, but is less obvious for long-chain n-alkanes.

These observed compositional differences are interpreted as weathering effects. The preferential removal of aromatic hydrocarbons has also been observed by Clayton and Swetland (1978) and may be interpreted to reflect the greater water solubility of aromatic hydrocarbons compared to saturated hydrocarbons, although the 2-3 times higher rate of aromatic versus saturated hydrocarbon depletion contrasts with the about 100 times greater water solubility (McAuliffe, 1966, see Tissot and Welte, 1984) of aromatic hydrocarbons of similar molecular weight. Also, data by Clayton and King (1987) on the concentration decrease of different aromatic compounds in weathered Phosphoria Shale do not support a simple solubility control. Data by Moss et al. (1988) and Schramedei (1991) indicate that oxidation rather than water washing or biodegradation is the principal process affecting organic matter during weathering.

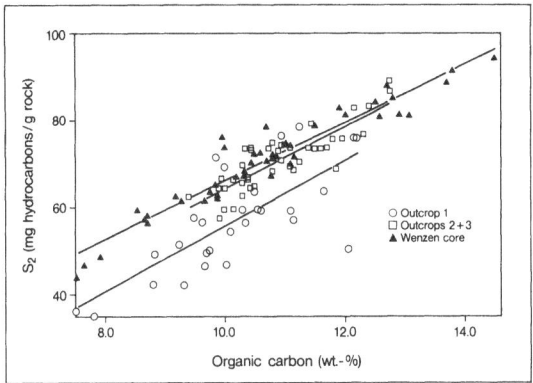

Fig. 79:   Correlation of organic carbon percentages and yield of hydrocarbon release upon Rock-Eval pyrolysis (S_2) for the Wenzen core, outcrops 2 and 3, and outcrop 1 (from Littke et al., 1991d).

The difference in yield of soluble organic matter between the two black shale sample series contrasts with the greater similarity in average thermal extract (S_1) yields obtained at 300°C (thermovaporization) prior to Rock-Eval pyrolysis (Table 31). However, thermovaporization-GC of selected severely weathered and unweathered samples revealed great compositional differences. For example, in the most strongly weathered samples, benzene is the by far most abundant single compound in the C_{5+}-range, whereas C_{12}-C_{18} n-alkanes and other high-molecular-weight compounds dominate in the unweathered samples. Therefore, the bitumen in the weathered samples is tentatively interpreted as a mixture of residual original bitumen, i.e., the rest of the bitumen which is also present in the core samples, and newly generated bitumen derived from the weathering-related kerogen degradation. Thus, S_1- and solvent extract-yields are not only influenced by loss of original bitumen, but also by addition of new products created during weathering. In the case of solvent extracts, loss is greater than addition, but in the case of thermal extracts loss and addition almost match each other. Possibly, during weathering low-molecular-weight products are preferentially added which are measured as thermovaporization products but not in solvent extracts. The assumption of a formation of new bitumen partly replacing the bitumen present in unweathered rocks implies that losses of the original bitumen are greater than directly evident from our data, i.e., more than 16% of the saturated hydrocarbons and more than 48% of the aromatic hydrocarbons present in unweathered Posidonia Shale were removed during weathering.

8.5   Rates of weathering processes

Calculation of weathering-related rates of loss for elements such as sulphur and organic carbon require 1) data on the mass loss from a given, unweathered (weight-)unit of rock, and 2) the timespan during which weathering affected and changed the rocks. If these data are available, the rate (R) of loss from a given volume of shale for an element (i) can be calculated as

$$R = \Delta_{rel}\, m_i \cdot m_{i1} \cdot \rho_r / t$$

where $\Delta_{rel}\, m_i$ is the relative loss due to weathering of the individual petrographic components, m_{i1} is the weight-percentage of these components in the unweathered rocks (see Table 29 and Appendix II), ρ_r is the average bulk dry rock density (kg/m^3) of unweathered rocks and t (a) is the time during which weathering was effective. R is expressed in units of kg of substance lost per cubicmetre of shale per year. The bulk dry rock density of Posidonia Shale cores is $2100\ kg/m^3$ (Mann and Müller, 1988).

In the case of the Hils syncline, data on the mass loss of sulphur and organic carbon are believed to be as accurate as possible for this geologic process due to the relative homogeneity of primary composition (and maturity) of the Posidonia Shale even over distances greater than about 150 m (Littke and Rullkötter, 1987) and due to the great number of samples analyzed. For the rate calculation, scenarios I and II were used (Fig. 78).

The second prerequisite regarding knowledge of the timespan of weathering is much more speculative. The first approach (A) will use the knowledge that weathering affects rocks to a depth of about 5 m in the Hils area, i.e., if erosion brings rocks to a depth of 5 m from the surface, weathering starts. It is also based on the calculated thickness of overburden (1450 m; Düppenbecker, in press) prior to erosion which started in the Middle Cretaceous (95 mabp). Under the assumption of a linear erosion rate, or an average erosion rate affecting the rocks during the Holocene, erosion of 1 m took about 65,500 a (years). For samples situated about 1.5 m below the top of the outcrop (vertically in the centre) a timespan of weathering of about 230,000 a (65,500 · 3.5) results.

In a second, less statistically oriented approach (B) it is assumed that weathering started only 10,000 abp (= years before present) after the last glacial period, the "Weichsel-Eiszeit" (10,000 - 70,000 abp) when glaciers almost reached the Elbe river close to the Hils area. The Scandinavian ice shield of several kilometres thickness started to melt at about 17,000 abp and was completely removed at about 8,000 abp (Woldstedt, 1958). Weathering-rates are known to be smaller at lower temperatures and it appears reasonable to believe that weathering was minor during the last glacial period. Thus, for the second approach, it is calculated with t = 10,000.

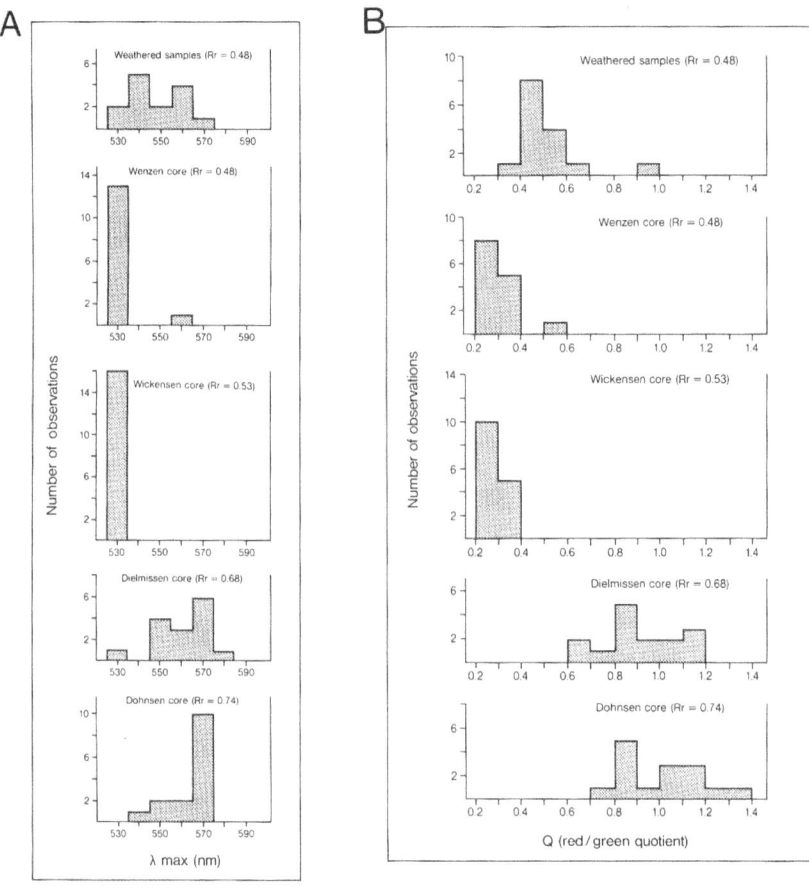

Fig. 80:   Histograms of fluorescence parameters measured on large tasmanales-alginite (alginite A) from the weathered outcrops and four different cores. λ_{max} values are wavelengths of maximum fluorescence intensity (Fig. A). Q-values (Fig. B) are quotients of fluorescence intensities at 650 and 500 nm (red/green quotient) and are commonly used as maturity parameter (Stach et al., 1982). Rr = mean random vitrinite reflectance values (from Littke et al., 1991d).

Compound	S1A	S1B	S2A	S2B
S	$2.3 \cdot 10^{-4}$	$5.2 \cdot 10^{-3}$	$2.3 \cdot 10^{-4}$	$5.3 \cdot 10^{-3}$
H_2SO_4	$7.0 \cdot 10^{-4}$	$1.5 \cdot 10^{-2}$	$7.0 \cdot 10^{-4}$	$1.6 \cdot 10^{-2}$
Org. Mat.	$1.4 \cdot 10^{-4}$	$3.3 \cdot 10^{-3}$	$2.1 \cdot 10^{-4}$	$4.8 \cdot 10^{-3}$

Table 32: Rates of sulphur, sulphuric acid, and organic matter release (kg m^{-3} a^{-1}) for scenarios I (S1) and II (S2) and weathering durations of 230,000 (A) and 10,000 (B) years (see Fig. 81).

The resulting rates and masses of sulphur, sulphuric acid and organic matter release are summarized in Table 32 and Fig. 81. The naturally released weathering products were transported into the hydrosphere. These natural processes add to the anthropogenic pollution of groundwater. Part of the weathering products probably reacted with minerals already within the Posidonia Shale, e.g. sulphuric acid probably enhanced calcite dissolution The apparently small rates of the natural processes compared to anthropogenic pollution (see Fig. 81) are partly compensated by the great volumes of rocks exposed to weathering conditions. For example, the outcrop of Posidonia Shale in the Hils syncline according to the map of Mann (1987) extends about 40 km in northwest-southeast direction with a width of 115 m on the surface; thus a total volume of $2.3 \cdot 10^7$ m^3 ($5 \cdot 115 \cdot 40,000$) of Posidonia Shale affected by weathering results. This outcrop is expected to release 3.22 - 110.4 t of organic matter per year and 16.1 - 368.0 t of sulphuric acid per year according to the calculated rates (Table 32). This number compares to 9.2 t S or 28.2 t H_2SO_4 (Fig. 76) deposited in the same area due to anthropogenic SO_2 emission if an emission rate of 2000 mg $S/m^2/a$ is assumed (Häfele et al., 1986: Fig. 11).

It should be noted that all calculations are only valid for processes in humid, temperate climate and that different climates as well as different rock fabrics will influence the rate of weathering.

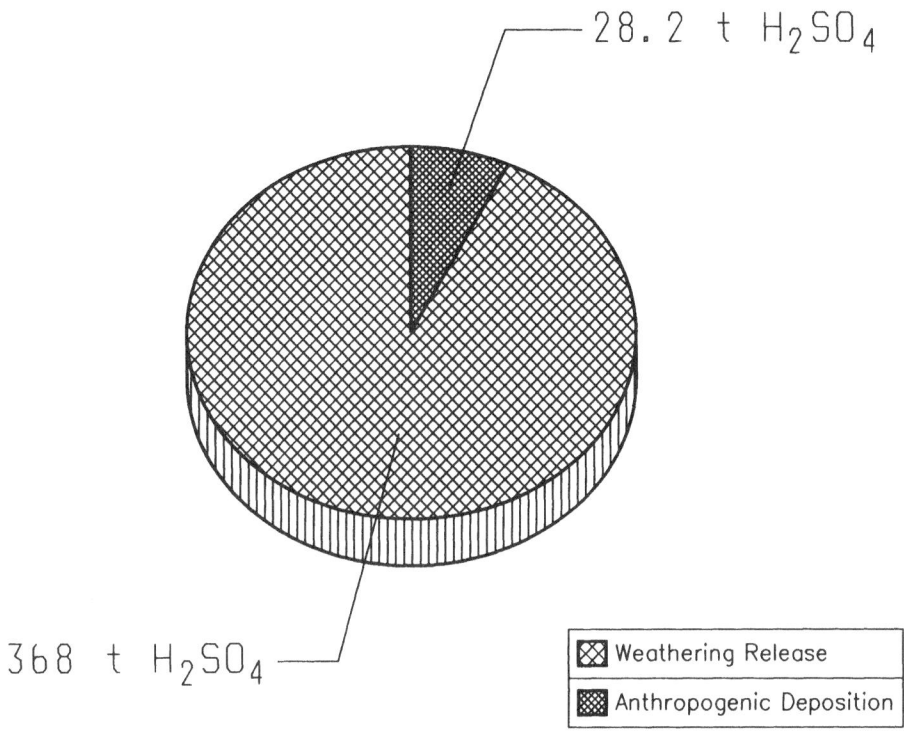

28.2 t H₂SO₄

368 t H₂SO₄

Weathering Release

Anthropogenic Deposition

Fig. 81:   Weathering related yearly H_2SO_4-release (368t) from the Posidonia Shale outcrop in the Hils area with the assumption of a weathering duration of 10,000a (see Table 32) compared to yearly anthropogenic H_2SO_4-deposition (28.2t) (see text).

9.   SUMMARY

9.1   Deposition of organic matter-rich sediments

The studies of recent and ancient organic matter-rich sediments presented in Chapter 3 reveal that the accumulation of organic matter depends on a variety of environmental parameters including bioproductivity, basin topography, current activity, and sedimentation rate.

The great influence of bioproductivity emphazised by Pedersen and Calvert (1990) becomes evident from the enrichment of organic matter in upwelling areas such as offshore Peru, northwest Africa, and Oman. Biologically produced organic matter is not only the direct source of particulate sedimentary organic matter, but it is also the most import sink for oxygen and other oxidizing agents in bottom and pore waters. The lack of these oxidizing agents enhances the preservation of organic matter.

The example of the Peruvian upwelling area does, however, also indicate that productivity of organic matter is not the sole control of organic matter content of sediments (Fig. 11), but that the sea floor topography and related current activity also play an important role in organic matter deposition.

With regard to productivity, it has further to be taken into account that the lability of the biosynthezised organic matter may have an effect on organic matter contents in sediments. The carbonate-rich, organic matter-poor limestones in the Posidonia Shale (Chapter 4) are regarded as indicators for times during which dominantely carbonate plankton was produced; obviously little of the related organic matter was preserved (Fig. 30) -probably due to its lability - although bottom waters remained anoxic.

Sedimentation rate is known as another parameter influencing organic matter accumulation and organic matter content in sediments. In oxic environments, not only organic matter accumulation, but also organic matter content correlate positively (with few exceptions) with sedimentation rate (Müller and Suess, 1979).

The above correlation is not valid for many anoxic settings in which high sedimentation rates are believed to have a diluting effect on organic matter content (e.g., in the case of the Black Sea). The diluting effect of high sedimentation rates becomes most obvious in the case of evaporitic sequences in which the most rapidly deposited rock salts contain the least organic matter, although accumulation rates are comparatively high. On the other hand, the Posidonia Shale is an example for the opposite effect, i.e., sedimentation rates correlate positively with organic matter contents as revealed by a comparison of the same ammonite zones in northern and southern Germany.

Like the amount, the petrographic and chemical composition of sedimentary organic matter depends on different environmental factors such as type of bioproduced organic matter, preservation conditions, and transport of organic particles.

In general, organic matter derived from higher land plants is less hydrogen-rich and less oil-prone than well preserved organic matter from aquatic organisms. This is documented by the lower hydrogen index (HI) values of coals and of terrestrial organic matter in fluvial and deltaic siltstones as compared to organic matter in lake sediments (compare Chapters 3.8 and 3.9).

In addition, the degree of degradation has a great impact on the chemical composition of the organic matter (e.g., Fig. 31) Strong degradation before and after deposition leads to hydrogen-poor sedimentary organic matter. An example is provided by the upwelling area offshore Oman (Fig. 10), where sediments at sites situated within the recent oxygen-minimum zone display higher HI-values than sediments drilled below the oxygen-minimum zone.

The petrographic composition of organic matter is not only influenced by the type of biosynthezised organic matter and preservation conditions, but also by the transport of organic particles. In general, large transport distances increase the inertinite/vitrinite ratio and lead to overall smaller grain sizes of organoclasts (Figs. 7 and 8).

Although relationships between environmental controls and features of sedimentary organic matter have been studied and elucidated quite extensively in the past decade, open questions still remain. Also, even detailed studies on ancient organic matter-rich sediments do not permit an undisputable reconstruction of depositional environments. This is because many important factors (rather than one or two) govern organic matter deposition.

9.2   Diagenesis of organic matter-rich sediments

Burial of sedimentary organic matter to great depths leads to the formation of oil and hydrocarbon gas. The controlling parameters of this transformation are temperature and the time during which specific temperatures are maintained as well as the thermal lability of the organic matter. The fact that controlling parameters are only few allowed the first numerical simulations of petroleum generation already ten to twenty years ago.

The degree of petroleum generation that sedimentary organic matter has experienced is commonly measured by optical and chemical maturity parameters. Many of the chemical maturity parameters are, however, strongly facies-dependent and can only be used for well-defined lithologies. The major weakness of vitrinite reflectance as the most widely used optical maturity parameter lies in the subjectivity of the selection of the vitrinite particles.

Vitrinite reflectance like other maturity parameters is used for the calibration of temperature histories of sedimentary rocks which are numerically simulated according to burial history and palaeo-heat flow reconstructions. Rather accurate temperature models can be achivied, if - in addition to vitrinite reflectance - temperature parameters such as homogenization temperatures of fluid inclusions or apatite fission track data are used (Chapter 5.6.2) which under favourable circumstances provide absolute temperatures for a narrow time interval in the past.

A further problem in the calibration of temperature history models is the existence of different methods to calculate vitrinite reflectance data. The different calculation methods may lead to important differences in predicted reflectance levels (Fig. 49). At the present time, the Easy % R_o method (Sweeney and Burham, 1990) seems to provide the most reliable calculated vitrinite reflectance values. Using this calibration method, the amount of eroded sediments (450 m) could be reconstructed for a gap in the sedimentary sequence of a well in the Styrian basin (Fig. 50). The reconstruction of temperature histories was undertaken intenseley during the past decades, because it provides the most important constraints for the calculation of the timing of oil and gas generation.

The total masses of oil and gas generated and expelled from the Posidonia Shale were predicted on the base of a mass balance. One major result of this first quantitative study on petroleum generation is that more than 50% of the original organic matter was lost from this rock due to expulsion of oil and gas (Fig. 41 and 42). For coals, roughly similar conversion rates are calculated, but the expelled products contain by far more carbon dioxide (Fig. 73 - 75).

The expulsion of oil from organic matter-rich sediments often occurs along zones of relatively high permeability. In the case of the Kimmeridge Clay from the Brae area, highly permeable silt layers exist within the source rock (Plate 2, D and E). Many clastic and carbonate sediments rich in organic matter are primarily impermeable to oil; in these cases expulsion follows fractures as in the case of the Posidonia Shale (Plate 2, A-C) and the Brown Limestone (Plate 2, F). In other carbonates, stylolithes are the most probable expulsion conduits (Plate 1, K and L).

In coals, pore throats are generally smaller than some of the larger oil molecules. A mass balance based on elemental compositions of coals of similar original composition revealed that only about 20% of the hydrogen present at 0.70% vitrinite reflectance is lost during maturation to 0.92% vitrinite reflectance. As the loss of hydrogen is much greater from clastic source rocks such as the Posidonia Shale, a much lower oil expulsion efficiency can be assumed for coals. The hydrogen left in the coals is at higher temperatures converted to gas, mainly methane. This is the principle reason why coals are very efficient gas source rocks (Chapter 7).

In reservoir rocks, organic matter can microscopically be preserved where oil was cracked to gas and a solid residue (pyrobitumen). Observations on Rotliegend sandstones from northern Germany indicate that oil emplacement finalized any cementation, i.e., pores partly filled with solid bitumen are still open and all other pores are completely filled with cement. The patchy character of the solid bitumen occurence suggests that previous oil only existed in small areas (channels) within the sandstones (Chapter 6.2).

9.3   Weathering of organic matter-rich sediments

Weathering of organic matter-rich sediments is regarded as an important process, because it changes the petrological and chemical character of these rocks. Data measured on organic matter in weathered samples may be misleading, if these data are assumed to be represensative of organic matter in the same source rock at depth.

Weathering of Posidonia Shale in the Hils syncline led to a release of about 82% of the original sulphur by pyrite oxydation, about 25% of the original calcite, and about 14% of the original organic matter. The residual organic matter is slightly depleted in hydrogen and bitumen, especially in aromatic hydrocarbons (by about 50%). The fluorescence colour of alginite particles is shifted to longer wavelengths. Porosity and pore size diameters are greater than in unweathered samples.

The macroscopic appearance of the Posidonia Shale seems to be ineffective as an evaluation tool for the geochemical effects of weathering. The best screening tool to differentiate between unweathered and weathered source rocks is presumably the microscopic appearance of pyrite.

If only outcrop samples are available for assessment of the petroleum generation capacity or maturity, the observations and changes described here should be considered and may help substantially in a more rigorous evaluation of source rock potential. Furthermore, rates of sulphuric acid and organic matter release were calculated. This estimate of natural release rates of "pollutant" products which is smaller for non-source rocks may be useful for the assessment of acceptable limits of anthropogenic release of hydrocarbons and sulphur products.

Appendix I: Speciation of Fe and S

1. Core samples (unweathered state)

All of the iron present in the unweathered reference samples (core samples) is assumed to occur as pyrite (FeS_2). The total amount of pyrite in the samples is calculated from the iron content (m_{Fe}) according to:

A(1) $m_{FeS_2} = m_{Fe} \cdot M_{FeS_2}/M_{Fe}$

Here m_i denotes the mass and M_i the molar mass of component i. The molar masses of the relevant elements are:

$$M_{Fe} = 55.85 \text{ g/mol}$$
$$M_S = 32.06 \text{ g/mol}$$

The mass of sulphur occurring as FeS_2 is obtained as:

A(2) $m_S(FeS_2) = m_{Fe} \cdot 2M_S/M_{Fe}$

And the mass of sulphur in organic matter (OM) is calculated as the difference of total sulphur content and pyrite sulphur:

A(3) $m_S(OM) = m_S - m_S(FeS_2)$

This amount is assumed to be unaffected by weathering and to remain constant. Numerically, the following values were obtained from this calculation:

m_{FeS_2} =   5.64 wt.% (pyrite content of unweathered core sample))

$m_S(OM)$ =   0.49 wt.% (sulphur content of organic matter of unweathered and weathered samples)

2. Outcrop samples (weathered state)

The total sulphur content of these samples is envisaged to be split up between organic matter (OM) and pyrite. The fraction of sulphur present in the OM ($m_S(OM)$) is the same as in the unweathered reference samples, because in both cases the OM content is the same (14.4%).
The amount of pyrite sulphur is obtained by difference according to:

A(4) $m_S(FeS_2) = m_S - m_S(OM)$

A(4) $m_S(FeS_2) = m_S - m_S(OM)$

and this determines the amount of pyrite iron according to:

A(5) $m_{Fe}(FeS_2) = m_S(FeS_2) \cdot M_{Fe}/2/M_S$

The pyrite content of the samples is then calculated as:

A(6) $m_{FeS_2} = m_{Fe}(FeS_2) \cdot M_{FeS_2}/M_{Fe}$

The iron not occurring as pyrite is assumed to be present as FeOOH and its amount is given by the difference between total iron content (m_{Fe}) as obtained from the elementary analysis, and pyrite iron:

A(7) $m_{Fe}(FeOOH) = m_{Fe} - m_{Fe}(FeS_2).$

Finally, the FeOOH content is calculated according to the relationship

A(8) $m_{FeOOH} = m_{Fe}(FeOOH) \cdot M_{FeOOH}/M_{Fe}$

with:

$$M_O = 16.00 \text{ g/mol}$$
$$M_H = 1.01 \text{ g/mol}$$

The numerical results of these calculations are given in Table 3.

Appendix II: Mass balance and Volume balance

1.   Mass balance calculation

Silicate as the supposedly most stable petrographic component is chosen as the base of reference for the comparative mass and volume balance calculations. Losses or addition of other components were calculated with respect to silicate. If compaction did not occur, and silicate was not affected by weathering, the amount of silicate per unit volume of shale is the same in the unweathered and the weathered state (scenario I). If removal of part of the silicate is taken into consideration (scenario II), then its "volume concentration" will decrease. Minor losses of silicate, as envisaged here, will not affect the stability of the rock fabric and the "no compaction" hypothesis can be maintained.

The contents of the individual petrographic components (i) present in the unweathered and weathered samples were normalized to the silicate content according to:

$$A(9) \qquad X_{i1} = \frac{m_{i1}}{m_{silicate1}} \quad \text{and} \quad X_{i2} = \frac{m_{i2}}{m_{silicate2}}$$

where the indices 1 and 2 refer to the unweathered and the weathered states respectively. For the unweathered state the i's correspond to carbonate, pyrite, organic matter, and silicate. For the weathered state, goethite (FeOOH) is also included.

The quantities of the individual petrographic components per unit volume are given by:

$$A(10) \qquad C_{i1} = \frac{m_{i1}}{V} = X_{i1} \frac{m_{silicate1}}{V} = X_{i1} \cdot C_{silicate1}$$

for the unweathered state and

$$A(11) \qquad C_{i2} = \frac{m_{i2}}{V} = X_{i2} \frac{m_{silicate2}}{V} = X_{i2} \cdot C_{silicate2}$$

for the weathered state. Scenario I assumes that no silicate is lost during weathering i.e.

$$A(12) \qquad C_{silicate2} = C_{silicate1}$$

The **relative losses** due to weathering of the individual petrographic components with respect to the unweathered state are then by definition:

$$A(13) \qquad \Delta_{rel} \, m_i \quad = \quad (C_{i1} - C_{i2})/C_{i1}$$
$$= \quad 1 - X_{i2}/X_{i1}$$

For the second scenario a loss of 5% of silicate per unit volume of shale during weathering was assumed. Thus,

$$A(14) \qquad C_{silicate2} = 0.95 \cdot C_{silicate1}$$

Correspondingly, the relative losses of the petrographic components are calculated according to:

$$A(15) \qquad \Delta_{rel} \, m_i = 1 - 0.95 \cdot X_{i2}/X_{i1}$$

The bulk volume of the rock samples (V_{bulk}) comprises the pore volume V_{pore} and the total volume of petrological components (V_{pet}).

If no compaction is assumed, the mass losses of petrological components due to weathering will result in an increase of the pore volume (porosity).

For the unweathered state (1) the total volume of the petrological components is given by:

A(16) $\qquad V_1 = \sum_i V_{i1} \qquad$ i = silicate, carbonate, pyrite, OM

The volume ratio of component i and silicate is:

A(17) $\qquad \dfrac{V_{i1}}{V_{silicate1}} = \dfrac{m_{i1}}{m_{silicate1}} \cdot \dfrac{\rho_{silicate}}{\rho_i}$

$\qquad\qquad\qquad = X_{i1} \cdot \rho_{silicate}/\rho_i$

Thus,

A(18) $\qquad V_1 = V_{silicate1} \cdot (1 + X_{carb.1} \cdot \rho_{silicate}/\rho_{carb.} + X_{pyrite1} \cdot \rho_{silicate}/\rho_{pyrite}$
$\qquad\qquad\qquad + X_{OM1} \cdot \rho_{silicate}/\rho_{OM})$

The corresponding relationship for the weathered state is given as:

A(19) $\qquad V_2 = V_{silicate2} \cdot (1 + X_{carb.2} \cdot \rho_{silicate}/\rho_{carb.} + X_{pyrite2} \cdot \rho_{silicate}/\rho_{pyrite}$
$\qquad\qquad\qquad + X_{OM2} \cdot \rho_{silicate}/\rho_{OM} + X_{FeOOH2} \cdot \rho_{silicate}/\rho_{FeOOH})$

The following relationships hold for the porosity (U) of the unweathered (1) and weathered (2) rock respectively:

A(20) $\qquad \Phi_1 = 1 - V_1/V_{bulk1}$

A(21) $\qquad \Phi_2 = 1 - V_2/V_{bulk2}$

With the above-mentioned assumption that no compaction occurred ($V_{bulk2} = V_{bulk1}$) and that the amount of silicate remained constant during weathering ($V_{silicate2} = V_{silicate1}$; scenario I), these latter equations can be combined to yield an expression for the calculation of the shale porosity after weathering:

A(22) $\qquad \Phi_2 = 1 - V_2/V_1 \cdot (1 - \Phi_1)$

For an assumed silicate loss of 5% during weathering (scenario II) the corresponding relationship is:

A(23) $\qquad \Phi_2 = 1 - 0.95 \cdot V_2/V_1 \cdot (1 - \Phi_1)$

References

Allan, J. and Douglas, A. (1977). Variations in the content and distribution of n-alkanes in a series of Carboniferous vitrinites and sporinites of bituminous rank. Geochim. Cosmochim. Acta, 41, 1223-1230.

Baker, D.R. (1962). Organic geochemistry of Cherokee Group in southeastern Kansas and northeastern Oklahoma. Amer. Assoc. Petr. Geol. Bull., 46, 1621-1642.

Barbé, A., Grimalt, J.O., Pueyo, J.J., and Albaigés, J. (1990). Characterization of model evaporitic environments through the study of lipid components. In: Durand, B. and Béhar, F. (eds): Advances in organic geochemistry 1989, Org. Geochem., 16, 815-828.

Barker, C.E. and Pawlewicz, M.J. (1986). The correlation of vitrinite with maximum temperature in humic kerogen. In: Buntebarth, G. and Stegena, L. (eds): Paleogeothermics, 79-93, Springer-Verlag, New York.

Baskin, D.K. and Peters, K.E. (1992). Early generation characteristics of a sulfur-rich Monterey kerogen. Amer. Assoc. Petr. Geol. Bull., 76, 1-13.

Battaerd, H.A.J. and Evans, D.G. (1979). An alternative representation of coal composition data. Fuel, 58, 105-108.

Baturin, G.N. (1983). Some unique sedimentological and geochemical features of deposits in coastal upwelling regions. In: Thiede, J. and Suess, E. (eds): Coastal upwelling, Part B: Sedimentary Records of Ancient Coastal Upwelling, 11-28, Plenum Press, New York.

Baudin, F., Dercourt, J., Herbin, J.P., and Lachkar, G. (1988). Le Lias supérieur de la zone ioniènne (Grèce): Une sédimentation riche en carbone organique. Compte Rendu Académie Science Paris Serie II, 307, 985-990.

Baudin F., Herbin, J.-P., Bassoulet, J.-P., Dercourt J., Lachkar, G., Manivit, H., and Renard, M. (1990). Distribution of organic matter during the Toarcian in the Mediterranean Tethys and Middle East. In: Huc, A.Y. (ed.): Deposition of Organic Facies, Amer. Assoc. Petr. Geol. Studies in Geol., 30, 73-92.

Bauer, W. (1991). Zur Beziehung von Lithofazies und Organofazies im Posidonienschiefer (Lias epsilon) SW-Deutschlands aufgrund organischer-geochemischer und organisch-petrologischer Untersuchungen, 112 p., Diplomarbeit, RWTH Aachen.

Béhar, F. and Vandenbroucke, M. (1986). Représentation chimique de la structure des kérogèns et des asphaltènes en fonction de leur origine et de leur degrée d' évolution. Rev. Inst. Franc. Pétr., 41, 173-188.

Beletskaya, S.N., Borova, G.H., Levi, S.Sh., and Belikova A.R. (1976). Changes in the distribution of hydrocarbons in the pore system of sedimentary rocks during recrystallization of carbonates and subsidence of sediments. In: Vassoevich, N.B. and Timofeyev, P.P. (eds): Study of organic matter in recent and ancient sediments, 155-162, Nauka, Moscow, 155-162 (in Russian).

Berger, W.H. (1989). Global maps of ocean productivity. In: Berger, W.H., Smetacek, V.S., and Wefer, G. (eds): Productivity of the ocean: present and past. Dahlem Workshop Reports, Life Sciences Research Report, 44, 429-456, John Wiley and Sons, Chichester.

Berger, W.H., Smetacek, V.S., Wefer, G. (1989). Ocean productivity and paleoproductivity -- An overview. In: Berger, W.H. et al. (eds): Productivity of the ocean: present and past. Dahlem Workshop Reports, Life Sciences Research Report, 44, 1-34, John Wiley and Sons, Chichester.

Berner, R.A. (1984). Sedimentary pyrite formation: an update. Geochim. Cosmochim. Acta, 48, 605-615.

Berner, R.A. and Lasaga, A.C. (1989). Simulation des geochemischen Kohlenstoffkreislaufs. Spektrum der Wiss., 5/89, 54-63.

Bertrand, R. (1990). Correlations between the reflectances of vitrinite, chitinozoans, graptolites, and scolecodonts. Org. Geochem., 15, 565-574.

Betz, D., Führer, F., Greiner, G., and Plein, E. (1987). Evolution of the Lower Saxony Basin. In: Ziegler, P.A. (ed.): Compressional intra-plate deformations in the alpine foreland. Tectonophysics, 137, 127-170.

Bharati, S. and Larter, S. (1991). Origin, evolution and petroleum potential of a Cambrian source rock: evidence from an organic petrographic study. In: Manning, D. (ed.): Organic geochemistry, advances and applications in energy and the natural environment, 141-143, Manchester Univ. Press, Manchester.

Bishop, J.K.B. (1989). Regional extremes in particulate matter composition and flux: effects on the chemistry of the ocean interior. In: Berger, W.H., Smetacek, V.S., and Wefer, G. (eds): Productivity of the ocean: present and past. Dahlem Workshop Reports, Life Sciences Research Report, 44, 117-138, John Wiley and Sons, Chichester.

Blanc-Valleron, M.-M., Gely, J.-P., Schuler, M., Dany, F., and Ansart, M. (1991). La matière organique associée aux évaporites de la base du sel IV (Oligocène inférieur) du bassin de Mulhouse (Alsace, France). Bull. Soc. Geol. France, 162,113-122.

Blob, A.K., Rullkötter, J., and Welte, D.H. (1988). Direct determination of the aliphatic carbon content of individual macerals in petroleum source rocks by near-infrared microspectroscopy. In: Mattavelli, L. and Novelli, L. (eds): Advances in organic geochemistry 1987, Org. Geochem., 13, 1073-1078.

Bostick, N.H. (1979). Microscopic measurements on the level of catagenesis of solid organic matter in sedimentary rocks to aid exploration for petroleum and to determine former burial temperatures a review. Soc. Econ. Palaeont. Miner. Spec. Publ., 26, 17-43.

Boudou, J.R., Pelet, R., and Letolle, R. (1984). A model of the diagenetic evolution of coaly sedimentary organic matter. Geochim. Cosmochim. Acta, 48, 1357-1362.

Brady, B.T. (1974). Theory of earthquakes. I. A scale independent theory of rock failure. Pageophys., 112, 701-725.

Bralower, T.J. and Thierstein, H.R. (1987). Organic carbon and metal accumulation in Holocene and mid-Cretaceous marine sediments: paleoceanographic significance. In: Brooks, J. and Fleet, A.J. (eds): Marine petroleum source rocks, Geol. Soc. Spec. Publ., 26, 345-369.

Brauckmann, F.J. (1984). Hochdiagenese im Muschelkalk der Massive von Bramsche und Vlotho. Bochumer geol. geotech. Arb., 14, 1-195.

Brauckmann, F.J. and Littke R. (1989). Organische Spuren in Rotliegend-Sedimenten NW-Deutschlands. Nachr. Deutsche Geol. Ges., 41, 17-18.

Bray, E.E. and Evans, E.D. (1961). Distribution of n-paraffins as a clue to recognition for source beds. Geochim. Cosmochim. Acta, 22, 2-15.

Brooks, J.D. and Smith, J.W. (1967). The diagenesis of plant lipids during the formation of coal, petroleum and natural gas - I. Changes in the n-paraffin hydrocarbons. Geochim. Cosmochim. Acta, 31, 2389-2397.

Brumsack, H.-J. (1991). Inorganic geochemistry of the German "Posidonia Shale": palaeoenvironmental consequences. In: Tyson, R.V. and Pearson, T.H. (eds): Modern and ancient continental shelf anoxia, Geol. Soc. Spec. Publ., 58, 353-362.

Buiskool Taxopeus, J.M.A. (1983). Selection criteria for the use of vitrinite reflectance as a maturity tool. In: Brooks, J. (ed.): Petroleum geochemistry and exploration of Europe, Geol. Soc. Spec. Publ., 12, 295-307.

Buntebarth, G. (1985). Das Temperaturgefälle im Dach des Bramscher Massivs aufgrund von Inkohlungsuntersuchungen im Karbon von Ibbenbühren. Fortschr. Geol. Rheinld. Westf., 33, 255-264.

Buntebarth, G., Grebe, H., Teichmüller M., and Teichmüller, R. (1979). Inkohlungsuntersuchungen in der Forschungsbohrung Urach 3 und ihre geothermische Interpretation. Fortschr. Geol. Reinld. Westf., 27, 183-200.

Burruss, R.C. (1989). Paleotemperatures from fluid inclusions: Advances in theory and technique. In: Naeser, N.D. and Mc Culloh, T.H. (eds): Thermal history of sedimentary basins - methods and case histories, 119-131, Springer-Verlag, Berlin.

Calvert, S.E. (1987). Oceanographic controls on the accumulation of organic matter in marine sediments. In: Brooks, J. and Fleet, A.J. (eds): Marine petroleum source rocks, 137-151, Blackwell Scientific, London.

Chandra, D. and Taylor, G.H. (1982). Gondwana coals. In: Stach, E., et al. (eds): Stach's textbook of coal petrology, 177-197, Gebr. Bornträger, Berlin.

Clayton, J.L. and King, J.D. (1987). Effects of weathering on biological marker and aromatic hydrocarbon composition of organic matter in Phosphoria shale outcrop. Geochim. Cosmochim. Acta, 51, 2153-2157.

Clayton, J.L. and Swetland, P.J. (1978). Subaerial weathering of sedimentary organic matter. Geochim. Cosmochim. Acta, 42, 305-312.

Cohen, Y., Krumbein, W.E., and Shilo, M. (1977). Solar Lake (Sinai). II. Distribution of photosynthtic microorganisms and primary production. Limmol. Oceanogr., 21, 609-620.

Cooles, G.P., Mackenzie, A.S., Quigley, T.M. (1986). Calculation of petroleum masses generated and expelled from source rocks. In: Leythaeuser, D. and Rullkötter, J. (eds): Advances in organic geochemistry 1985, Org. Geochem., 10, 235-245.

Curiale, J.A. (1986). Origin of solid bitumens, with emphasis on biological marker results. In: Leythaeuser, D. and Rullkötter, J. (eds): Advances in organic geochemistry 1985, Org. Geochem., 10, 559-580.

Daumas, R., Laborde, P., Paul, R., Romano, J.C. and Sautriot, D. (1978). Les méchanismes de transformation de la matière organique en Atlantique intertropical. Étude de la mineralisation et de la diagenèse dans les sédiments superficiels. In: Combaz, A. and Pelet, R., (eds) ORGON II, Geochémie Organique des Sédiments Marins Profonds, 43-88, Editions du CNRS, Paris.

Dean, W.E. and Arthur, M.A. (1989). Iron-sulfur-carbon relationships in organic-carbon-rich sequences. I: Cretaceous Western Interior seaway. Amer. Journ. of Science, 289, 708-743.

Debelak, K.A. and Schrodt, J.T. (1979). Comparison of pore structure in Kentucky coals by mercury penetration and carbon dioxide adsorption. Fuel, 58, 732-736.

Degens, E.T., Emeis, K.-C., Mycke, B., and Wiesner, M.G. (1986). Turbidites: the principle mechanism yielding black shales in the early deep Atlantic ocean. Geol. Soc. Spec. Publ., 21, 361-376.

Degens, E.T. and Ittekott, V. (1987). The carbon cycle-tracking the path of organic particles from sea to sediment. In: Brooks, J. and Fleet, A.J. (eds): Marine petroleum source rocks, Geol. Soc. Spec. Publ., 26, 121-136.

Degens E.T. and Mopper, K. (1976). Factors controlling the distribution and early diagenesis of organic material in marine sediments. In: Riley, J.P. and Chester, R. (eds): Chemical Oceanography, 6, 60-114.

Degens, E.T., Stoffers, P., Golubic, S., and Dickman, M.D. (1978). Varve chronology: estimated rates of sedimentation in the Black Sea deep basin. In: Ross, D.A., Neprochnor, Y.P., et al. (eds): Initial Reports Deep Sea Drilling Project, 42, 499-508.

Demaison, G. (1991). Anoxia vs productivity: what controls the formation of organic-carbon-rich sediments and sedimentary rocks?: Discussion. Amer. Assoc. Petr. Geol. Bull., 75, 499.

Demaison, G.J. and Huizinga, B.J. (1991). Genetic classification of petroleum systems. Amer. Assoc. Petr. Geol. Bull., 75, 1626-1643.

Demaison, G.J. and Moore, G.T. (1980). Anoxic environments and oil source bed genesis. Amer. Assoc. Petr. Geol. Bull., 64, 1179-1209.

Deuser, W.G. (1974). Evolution of anoxic conditions in Black Sea during Holocene. In: Degens, E.T. and Ross, D.A. (eds): The Black Sea - geology, chemistry, and biology. Amer. Assoc. Petr. Geol. Memoir, 20, 133-136.

Diessel, C.F.K. (1987). On the correlation between coal facies and depositional environments. Proc. 20th Symp. Adv. in the Study of the Sydney Basin, 19-22, Newcastle.

Düppenbecker, S.J. (1992). Genese und Expulsion von Kohlenwasserstoffen in zwei Regionen des Niedersächsischen Beckens unter besonderer Berücksichtigung der Aufheizraten. Ber. Forschungszentrum Jülich, 2657, 1-304.

Düppenbecker, S.J., Dohmen, L., and Welte, D.H. (1991). Numerical modelling of petroleum expulsion in the Lower Saxony basin (NW-Germany). In: England, W.A. and Fleet, A.J. (eds): Petroleum migration, Geol. Soc. Spec. Publ., 59, 47-64.

Durand, B. (1980). Kerogen. Insoluble organic matter from sedimentary rocks, 519 p., Editions Technip., Paris.

Durand, B. (1988). Understanding of HC migration in sedimentary basins (present state of the knowledge). In: Mattavelli, L. and Novelli, L. (eds): Advances in organic geochemistry 1987, Org. Geochem., 13, 445-459.

Durand, B., Espitalié, J., Nicaise, G., and Combaz, A. (1972). Études de la matière organique insoluble (kérogène) des argiles du Toarcian. Analyse élémentaire, études en microscopie et diffraction électroniques. Revue Inst. Franç. Pétr., 27, 865-884.

Durand, B. and Paratte, M. (1983). Oil potential of coals. In: Brooks, J. (ed.): Petroleum geochemistry and exploration of Europe, 285-292, Blackwell Scientific Publications, Oxford.

Ebner, F. and Sachsenhofer, R.F. (1991). Die Entwicklungsgeschichte des Steirischen Tertiärbeckens. Mitt. Abt. Geol. und Paläont. Landesmuseum Joanneum, 49, 1-96.

Emeis, K.-C. and Morse, J.W. (1990). Organic carbon, reduced sulfur and iron relationships in sediments of the Peru margin, Sites 680 and 688. In: Suess, E., von Huene, R., et al. (eds) Proceedings of the Ocean Drilling Program, Sci. Results, 112, 441-454.

Emeis, K.C., Morse, J.W., and Mays, L.L. (1991). Organic carbon, reduced sulfur, and iron in Miocene to Holocene upwelling sediments from the Oman and Benguela upwelling systems. In: Prell, W.L., Niitsuma, N., et al. (eds): Proceed. ODP, Sci. Results, 117, 517-528, College Station, TX (Ocean Drilling Program).

von Engelhardt, W. (1973). Sedimentpetrologie, Teil III: Die Bildung von Sedimenten und Sedimentgesteinen. 378p., Schweizerbarth, Stuttgart.

England, W.A., Mackenzie, A.S., Mann, D.M., and Quigley, T.M. 1987: The movement and entrapment of petroleum fluids in the subsurface. Journ. Geol. Soc. London, 144, 327-347.

Epstein, A.G., Epstein, J.B., and Harris, L.D. (1977). Conodont colour alteration - an index to organic metamorphism. U.S. Geol. Surv. Prof. Pap., 995, 1-27.

Espitalié, J., Laporte, J.L., Madec, M., Marquis, F., Leplat, P., Paulet, J., and Boutefeu, A. (1977). Méthode rapide de caractérisation des roches mères, de leur potentiel pétrolier et de leur degré d' évolution. Rev. Inst. Franç. Pétr., 32, 23-42.

Espitalié, J., Marquis, F., Sage, L., and Barsony, I. (1987). Géochemie organique du basin de Paris, Revue Inst. Franç. Pétr., 42, 271-302.

Espitalié, J., Maxwell, J.R., Chenet, Y., and Marquis, F. (1988). Aspects of hydrocarbon migration in the Mesozoic in the Paris basin as deducted from an organic geochemical survey. In: Mattavelli, L. and Novelli, L. (eds): Advances in organic geochemistry 1987, Org. Geochem., 13, 467-482.

Ferguson, J. (1987). The significance of carbonate ooids in petroleum source rock studies. In: Brooks, J. and Fleet, A.J. (eds): Marine petroleum source rocks, Geol. Soc. Spec. Publ., 26, 207-215.

Ferguson, J. and Ibe, A.C. (1982). Some aspects of the occurrence of proto-kerogen in Recent ooids. Jour. Petr. Geol., 4, 267-285.

Fisher, D. St. J. and Hudson, J.D. (1987). Pyrite formation in Jurassic shales of contrasting biofacies. In: Brooks, J. and Fleet, A.J. (eds): Marine petroleum source rocks, Geol. Soc. Spec. Publ., 26, 69-78.

Fleet, A.J., Clayton, C.J., Jenkyns, H.C., and Parkinson, D.N. (1987). Liassic source-rock deposition in Western Europe. In: Brooks, J. and Glennie, K.W. (eds): Petroleum geology of North West Europe, 59-70, Graham and Trotmann, London.

Flügel, H.W. (1988). Geologische Karte des prätertiären Untergrundes. In: Kröll, A. et al.: Erläuterungen zu den Karten über den prätertiären Untergrund des Steirischen Beckens und der Südburgenländischen Schwelle, 21-43, Wien (Geol. B.-A.).

Forsberg, A. and Bjoréy, M. (1981). Sedimentological and organic geochemical study of the Botneheia Formation, Svalbard, with special emphasis on the effects of weathering on the organic matter in shales. Org. Geochem., 5, 60-68.

Franzen, J.L. and Michaelis, W. (1988, eds). Der eozäne Messelsee - Eocene Lake Messel. Cour. Forsch.-Inst. Senckenberg, 107, 452p.

Galloway, W.E. and Hobday, D.K. (1983). Terrigenous clastic depositional systems. Applications to petroleum, coal and uranium exploration. 423p., Springer, New York.

Geyer, O.F. and Gwinner, M.P. (1962). Der schwäbische Jura. 472 p., Schweizerbarth, Stuttgart.

Given, P.H. (1984). An essay on the organic geochemistry of coal. In: Gorbaty, M.L., Larsen, J.W., and Wender, J. (eds): Coal Science, 3, 65-253 + 339-341, Academic Press, Orlando.

Glenn, C.R. and Arthur, M.A. (1984). Sedimentary and geochemical indicators of productivity and oxygen contents in modern and ancient basins: the Holocene Black Sea as the "type" anoxic basin. Chemical Geology, 48, 325-354.

Goodarzi, F., Fowler, M.G., Bustin, M., and Mc Kirdy, D.M. (1992). Thermal maturity of early Paleozoic sediments as determined by the optical properties of marine-derived organic matter - a review. In: Schidlowski, M. et al. (eds): Early organic evolution: Implications for mineral and energy resources, 279-295, Springer-Verlag, Berlin.

Goossens, H., de Lange, F., de Leeuw, J.W., and Schenck, P.A. (1988). The pristane formation index, a molecular maturity parameter. Confirmation in samples from the Paris basin. Geochim. Cosmochim. Acta, 52, 2439-2444.

Goossens, H., Rijpstra, W.I.C., Dueren, R.R., de Leeuw, J.W., and Schenck, P.A. (1986). Bacterial contribution to sedimentary organic matter; a comparative study of lipid moieties in bacteria and recent sediments. In: Leythaeuser, D. and Rullkötter, J. (eds): Advances in organic geochemistry 1986, Org. Geochem., 10, 683 696.

Gutjahr, C.C.M. (1966). Carbonization measurements of pollen grains and spores and their application. Leidse Geol. Meded., 38, 1-29.

Gutjahr, C.M.M. (1983). Introduction to incident light microscopy of oil and gas source rocks. Geol. Mijnbouw, 62, 417-425.

Häfele, W., Riemer, H., Ausubel, J.H. and Geiss, H. (1986). Future uses of fossil fuels: A global view of related emissions and depositions. In: Leythaeuser, D. and Rullkötter, J. (eds): Advances in organic geochemistry 1985, Org. Geochem. 10, 1-15.

Hagemann, H. and Hollerbach, A. (1981). Spectral fluorometric analysis of extracts, a new method for the determination of the degree of maturity of organic matter in sedimentary rocks. Bull. Centres Rech. Expl.-Prod. Elf-Aquitaines, 5, 635-650.

Hahn A., Kind E.G., and Mishra D.L. (1976). Depth estimation of magnetic sources by means of Fourrier amplitude spectra. Geophys. Prospecting, 24, 287-308.

Hallam, A. (1967). An environmental study of the Upper Domerian and Lower Toarcian in Great Britain. Philos. Trans. Royal Soc. London, 252, 393-445.

Hallam, A. (1975). Jurassic Environments. 269p., Cambr. Univ. Press, New York.

Hallam, A. (1981). A revised sea-level curve for the early Jurassic. Journ. Geol. Soc. London, 138, 735-743.

Hallam, A. (1981). Facies interpretation and the stratigraphic record. 291 p., W.H. Freeman, Oxford.

Harland, W.B., Armstrong, R.L., Cox, A.V., Craig, L.E., Smith, A.G., and Smith, D.G. (1990). A geologic time scale 1989, 263p., Cambr. Univ. Press, New York.

Harris, L.A. and Yust, C.S. (1976). Transmission electron microscope observations of porosity in coal. Fuel, 55, 233-236.

Harris, E.C. Jr. and Petersen, E.E. (1979). Change in physical characteristics of Roland seam coal with progressive solvent extraction. Fuel, 58, 599-602.

Harvey, H.R., Fallon, R.D., and Patton, J.S. (1986). The effect of organic matter and oxygen on the degradation of bacterial membrane lipids in marine sediments. Geochim. Cosmochim. Acta, 50, 795-804.

ten Haven, H.L., de Leeuw, J.W., Rullkötter, J., and Sinninghe-Damsté, J.S. (1988). Restricted utility of the pristane/phytane ratio as palaeoenvironmental indicator. Nature, 330, 641-643.

ten Haven, H.L., Littke, R., Rullkötter, J., Stein, R., and Welte, D.H. (1990). Accumulation rates and composition of organic matter in Late Cenozoic sediments underlying the active upwelling area off Peru. In: Suess, E., von Huene, R., et al. (eds): Proceedings of the Ocean Drilling Program, Sci. Results, 112, 591-605.

ten Haven, H.L., Littke R., and Rullkötter, J. (1992). Hydrocarbon biological markers in Carboniferous coals of different maturities. In: Moldowan, J.M., Albrecht, P., and Philp, R.P. (eds): Biological markers in sediments and petroleum, 142-155, Prentice Hall, Englewood Cliffs, New Jersey.

ten Haven, H.L. and Rullkötter, J. (1991). Preliminary lipid analysis of sediments recovered during Leg 117. In: Prell, W.L., Niitsuma, N., et al. (eds): Proceedings of the Ocean Drilling Program, Sci. Results, 117, 561-570.

Heckel, P.H. (1991). Thin widespread Pennsylvanian black shales of Midcontinent North America: a record of a cyclic succession of widespread pycnoclines in a fluctuating epeiric sea. In: Tyson, R.V. and Pearson, T.H. (eds): Modern and ancient continental shelf anoxia, Geol. Soc. Spec. Publ., 58, 259-274.

Heckers, J. (in press). Organisch-geochemische und karbonatpetrographische Untersuchungen zur Entstehung des Asphaltkalkvorkommens von Holzen/Ith. Diss., Univ. Marburg.

Hite, R.J. and Anders D.E. (1991). Petroleum and evaporites. In: Melvin, J.L. (ed.): Evaporites, petroleum and mineral resources. Developm. in Sedimentology, 50, 349-410.

Hofmann, P. (1992). Sedimentary facies, organic facies and hydrocarbon generation in evaporitic sediments of the Mulhouse basin, France. Ber. Forschungszentrum Jülich, 2664, 1-288.

Hollerbach, A. and Hagemann, H.W. (1981). Organic geochemical and petrological investigations into a series of coals with increasing rank. Proceed. Intern. Conference on Coal Science, 80-85, Glückauf Verlag, Essen.

Horsfield, B. (1989). Practical criteria for classifying kerogens: some observations from pyrolysis-gas chromatography. Geochim. Cosmochim. Acta, 53, 891-901.

Horsfield, B., Bharati, S., Larter, S.R., Leistner, F., Littke, R., Schenk, H.J., and Dypvik, H. (1992). On the atypical petroleum-generating characteristics of alginite in the Cambrian Alum Shale. In: Schidlowski, M. et al. (eds): Early organic evolution. Implications for mineral and energy resources, 257-266, Springer-Verlag, Berlin.

Horsfield, B., Heckers, J., Leythaeuser, D., Littke, R., and Mann, U. (1991). A study of the Holzener Asphaltkalk, northern Germany: Observations regarding the distribution, composition and origin of organic matter in an exhumed petroleum reservoir. Mar. Petrol. Geol., 8, 198-211.

Horsfield, B., Yordy, K.L., and Crelling, J.C. (1988). Determining the petroleum-generating potential of coal using organic geochemistry and organic petrology. In: Mattavelli, L. and Novelli, L. (eds): Advances in organic geochemistry 1987, Org. Geochem., 13, 121-129.

Horvath, F., Dövenyi, P., Szalay, A., and Royden, L.H. (1988). Subsidence, thermal, and maturation history of the Great Hungarian Plain. In: Royden, L.H. and Horvath, F. (eds): The Pannonian basin. A study in basin evolution, Amer. Assoc. Petr. Geol. Mem., 45, 355-372.

Huc, A.Y. (1977). Contribution de la geochimie organique a une esquisse palaeoécologique des schistes bitumineux de Toarcien de l'est du bassin de Paris. Étude de la matière organique insoluble (kérogène). Rev. Inst. Franç. Pétr., 32, 703-718.

Huc, A.Y. (1988a). Aspects of depositional processes of organic matter in sedimentary basins. In: Mattavelli, L. and Novelli, L. (eds): Advances in organic geochemistry 1987, Org. Geochem., 13, 263-272.

Huc, A.Y. (1988b). Sedimentology of organic matter. In: Frimmel, F.H. and Christman, R.F. (eds): Humic substances and their role in the environment, 215-243, John Wiley and Sons, Chichester.

Huc, A.Y. (1990, ed.). Deposition of organic facies. Amer. Assoc. Petr. Geol., Studies in Geol., 30 , 41-56.

Huc, A.Y., Durand, B., Roucachet, J., Vandenbroucke, M., and Pittion, J.C. (1986). Comparison of three series of organic matter of continental origin. In: Leythaeuser, D. and Rullkötter, J. (eds): Advances in organic geochemistry 1985, Org. Geochem., 10, 65-73.

Hunt, J. (1979). Petroleum geochemistry and geology. W.H. Freeman and Company, San Francisco, 617p.

Hunt, J. (1991). Generation of gas and oil from coal and other terrestrial organic matter. Org. Geochem., 17, 673-680.

Hutton, A.C. and Cook, A.C. (1980). Influence of alginite on the reflectance of vitrinite from Joadja, NSW, and some other coals and oil shales containing alginite, Fuel, 59, 711-714.

Hutton, A.C., Kantsler, A.J., Cook, A.C., and McKirdy, D.M. (1980). Organic matter in oil shales. Austr. Petr. Expl. Assoc., 20, 44-67.

Hvoslef, S., Larter, S.R., and Leythaeuser, D. (1988). Aspects of generation and migration of hydrocarbons from coal-bearing strata of the Hitra Formation, Haltenbanken area, offshore Norway. In: Mattavelli, L. and Novelli, L. (eds): Advances in organic geochemistry 1987, Org. Geochem., 13, 525-536.

Jacob, H. (1961). Über bituminöse Schiefer, humose Tone, Brandschiefer und ähnliche Gesteine. Ein Beitrag zur Frage der Erdölgenesis aus kohlenpetrographischer Sicht. Erdöl Kohle, 14, 2-11.

Jacob, H. (1989). Classification, structure, genesis and practical importance of natural solid oil bitumen ("migrabitumen"). Int. Journ. Coal. Geol., 11, 65-79.

Janicke, A. (1990). Sedimentologie, Mineralogie und Geochemie des nordwestdeutschen Posidonienschiefers (Toarcium), Diss., 148 p., Univ. Hannover.

Jankowski, B. (1981). Die Geschichte der Sedimentation in Nördlinger Ries und Randecker Maar. Boch. geol. geotechn. Arb., 6, 1-315.

Jankowski, B. and Littke, R. (1986). Das organische Material der Ölschiefer von Messel. Geowissenschaften in unserer Zeit, 4, 73-80.

Janowsky, U. (1984). Experimentelle Untersuchungen zum Strömungs- und Sorptionsverhalten von Wasser und Gasen in Steinkohle und Ableitung eines Porenmodells. Diss., 144p., University of Essen, Essen.

Jenkyns, H.C. (1985). The early Toarcian and Cenomanian - Turonian anoxic events in Europe - comparisons and contrasts. Geol. Rundschau, 74, 505-518.

Jenkyns, H.C. (1988). The early Toarcian (Jurassic) anoxic event: stratigraphic, sedimentary and geochemical evidence. Amer. Journ. of Science, 288, 101-151.

Jochum, J. (in press). Intraformationaler Stofftransport von Karbonat, Haupt- und Spurenelementen sowie Erdöl-Kohlenwasserstoffen in Posidonienschieferprofilen (Hilsmulde) mit zunehmener Temperaturbeanspruchung. Diss., RWTH Aachen.

Jochum, J., Leythaeuser, D., Littke, R., and Ropertz, B. (1991). Oil-bearing fluid inclusions in calcite-filled horizontal fractures from mature Posidonia Shale (Hils syncline, NW-Germany). In: Manning, D.A.C. (ed.): Organic geochemistry. Advances and applications in energy and the natural environment, 160-162, Manchester Univ. Press, Manchester.

Jones, R.W. (1980). Some mass balance and geological constraints on migration mechanisms. In: Roberts, W.H., III and Cordell, R.J. (eds): Problems of petroleum migration. Amer. Assoc. Petr. Geol., Studies in Geology, 10, 47-68.

Jordan, R. (1974). Salz und Erdöl/Erdgas-Austritt als Fazies-bestimmende Faktoren im Mesozoikum Nordwest-Deutschlands. Geol. Jb. A, 13, 1-64.

Jüntgen, H. (1986). Wissenschaftliche Methoden zur Bestimmung der Eigenschaften von Kohlen in tiefliegenden Flözen. Erdöl und Kohle, Erdgas, Petrochemie, 39, 32-45.

Jüntgen, H. and Karweil, J. (1962). Künstliche Inkohlung von Steinkohlen. Freiberger Forsch.-Hefte, A229, 27-36.

Jüntgen, H. and Karweil, J. (1966a). Gasbildung und Gasspeicherung in Steinkohleflözen. I. Gasbildung. Erdöl und Kohle, Erdgas, Petrochemie, 19, 251-238.

Jüntgen, H. and Karweil, J. (1966b). Gasbildung und Gasspeicherung in Steinkohleflözen. II. Gasspeicherung. Erdöl und Kohle, Erdgas, Petrochemie, 19, 339-344.

Jüntgen, H. and Klein, J. (1975). Entstehung von Erdgas aus kohligen Sedimenten. Erdöl und Kohle, Erdgas, Petrochemie, 28, 65-73.

Kalkreuth, W., Steller, M., Wieschenkämper, I., and Ganz, S. (1991). Petrographic and chemical characterization of Canadian and German coals in relation to utilization potential. 1. Petrographic and chemical characterization of feedcoals. Fuel, 70, 683-694.

Karweil, J. (1956). Die Metamorphose der Kohlen vom Standpunkt der physikalischen Chemie. Z. Dtsch. Geol. Ges., 107, 132-139.

Katz, B.J., Kelley, P.A., Royle, R.A., and Jorjorian, T. (1991). Hydrocarbon products of coals as revealed by pyrolysis-gas chromatography. Org. Geochem., 17, 711-722.

Kauffman, E.G. (1979). Benthic environments and palaeoecology of the Posidonienschiefer (Toarcian). Neues Jb. Geol. Paläont. Abhandlungen, 157, 18-36.

Kauffman, E.G. (1981). Ecological reappraisal of the German Posidonienschiefer (Toarcian) and the stagnant basin model. In: Gray, J., Boucot, A.J., and Berry, W.B.N. (eds): Communities of the Past, 311-381, Hutchinson Ross, Stroudsburg.

Kenig, F. and Huc, A.Y. (1990). Incorporation of sulfur into recent organic matter in a carbonate environment (Abu Dhabi, United Arab Emirates). In: Orr, W.L. and White, C.M. (eds): Geochemistry of sulfur in fossil fuels, ACS Symposium Series, 429, 170-185.

Kenig, F., Huc, A.Y., Purser, B.H., and Oudin, J.-L. (1990). Sedimentation, distribution and diagenesis of organic matter in a recent carbonate environment, Abu Dhabi, U.A.E. In: Durand, B. and Béhar, F. (eds): Advances in organic geochemistry 1989, Org. Geochem., 16, 735-747.

Kennicutt II, M.C., De Freitas, D.A., Joyce, J.E., and Brooks, J.M. (1986). Nonvolatile organic matter in sediments from Sites 614 to 623, Deep Sea Drilling Project Leg 96. In: Bouma, A.H., Coleman, J.M., Meyer, A.W., et al. (eds): Init. Repts. DSDP, 96, 747-756.

Khavari-Khorasani, G. (1987). Oil-prone coals of the Walloon coal measures, Surat Basin/Australia. In: Scott, A.L. (ed.): Coal and coal-bearing strata: recent advances, Geol. Soc. Spec. Publ., 32, 303-310.

Koch, J. and Arnemann H. (1975). Die Inkohlung in Gesteinen des Rhät und Lias im südlichen Nordwestdeutschland. Geol. Jb., A29, 45-55.

Königshof, P. (1992). Der Farbänderungsindex von Conodonten (CAI) in paläozoischen Gesteinen (Mitteldevon bis Unterkarbon) des Rheinischen Schiefergebirges - eine Ergänzung zur Vitrinitreflexion. Courier Forsch.-Inst. Senckenberg, 1-112.

van Krevelen, D.W. (1961). Coal. Typology-Chemistry-Physics-Constitution. 513 p., Elsevier, Amsterdam.

Krooss, B.M., Leythaeuser, D., and Lillack, H. (in press). Nitrogen-rich natural gases: Qualitative and quantitative aspects of gas accumulation in reservoirs. Erdöl und Kohle, Erdgas, Petrochemie/Hydrocarbon Technology.

Kruijs, E. and Barron, E.J. (1990). Climate model prediction of paleoproductivity and potential source-rock distribution. In: Huc, A.Y. (ed.): Deposition of organic facies. Amer. Assoc. Petr. Geol. Studies in Geology, 30, 195-216.

Küspert, W. (1982). Environmental changes during oil shale deposition as deduced from stable isotope ratios. In: Einsele, G. and Seilacher, A. (eds): Cyclic and Event Stratification, 482-501, Springer-Verlag, Berlin.

Larter, S.R. (1989). Chemical models of vitrinite reflection evolution. Geol. Rundschau, 78, 349-359.

Larter, S.R., and Senftle, J.T. (1985). Improved kerogen typing for petroleum source rock analysis. Nature, 318, 277-280.

Leischner, K. (in press). Kalibration simulierter Temperaturgeschichten von Gesteinen mit organischen Reifeparametern und anorganischen Temperaturindikatoren. Diss. Univ. Bochum.

Leischner, K., Welte, D.H., and Littke, R. (1993). Fluid inclusions and organic maturity parameters as calibration tools in basin modelling. In: Doré, A.G., Augustson, J.H., Hermanrud, C., Steward, D.J., and Sylta, é. (eds): Basin Modelling: Advances and Applications, NPF Spec. Publ., 3, 161-172, Elsevier, Amsterdam.

Lerche, I. and McKenna, T. (1991). Pollen translucency as a thermal maturation indicator. Journ. Petrol. Geol., 14, 19-36.

de Leeuw, J.W. and Sinninghe Damsté, J.S. (1990). Organic sulfur compounds and other biomarkers as indicators of palaeosalinity. In: Orr, W.L. and White, C.M. (eds): Geochemistry of sulfur in fossil fuels, ACS Symposium Series, 429, 417-443.

Leventhal, J.S. (1983). An interpretation of the carbon and sulphur relationships in the Black Sea sediments as indicators of the environment of deposition. Geochim. Cosmochim. Acta, 47, 133-137.

Leythaeuser, D. (1973). Effects of weathering on organic matter in shales. Geochim. Cosmochim. Acta, 37, 113-120.

Leythaeuser, D., Littke, R., Radke, M., and Schaefer, R.G. (1988a). Effects of primary migration and petroleum expulsion recognized by geochemical analyses. In: Mattavelli, L. and Novelli, L. (eds): Advances in organic geochemistry 1987, Org. Geochem., 13, 489-502.

Leythaeuser, D., Radke, M., and Willsch, H. (1988c). Geochemical effects of primary migration of petroleum in Kimmeridge source rocks from Brae field area, North Sea. II: Molecular composition of alkylated naphthalenes, phenanthrenes, benzo- and dibenzothiophenes. Geochim. Cosmochim. Acta, 52, 2879-2891.

Leythaeuser, D., Schaefer, R.G., and Radke, M. (1988b). Geochemical effects of primary migration of petroleum in Kimmeridge source rocks from Brae field area, North Sea. I: Gross composition of C15+-soluble organic matter and molecular composition of C15+-saturated hydrocarbons. Geochim. Cosmochim. Acta, 52, 701-713.

Leythaeuser, D. and Welte, D.H. (1969). Relation between distribution of heavy n-paraffins and coalification in Carboniferous coals from the Saar district, Germany. In: Schenck, P.A. and Havenaar, I. (eds): Advances in organic geochemistry 1968, 429-442, Pergamon Press, Oxford.

Lin, R. and Davis, A. (1988). A fluorogeochemical model for coal macerals. Org. Geochem., 12, 363-374.

Littke, R. (1985). Flözaufbau in den Dorstener, Horster und Essener Schichten der Bohrung Wulfener Heide 1 (nördliches Ruhrgebiet). Fortschr. Geol. Rheinld. Westf., 33, 129-159.

Littke, R. (1987). Petrology and genesis of Upper Carboniferous coal seams from the Ruhr region, Western Germany. Int. Journ. Coal. Geol., 7, 147-184.

Littke, R., Baker, D.R., and Leythaeuser, D. (1988). Microscopic and sedimentologic evidence for the generation and migration of hydrocarbons in Toarcian source rocks of different maturities. In: Mattavelli, L. and Novelli, L. (eds): Advances in organic geochemistry 1987, Org. Geochem., 13, 549-559.

Littke, R., Baker, D.R., Leythaeuser, D., and Rullkötter, J. (1991b). Keys to the depositional history of the Posidonia Shale (Toarcian) in the Hils syncline, northern Germany. In: Tyson, R.V. and Pearson, T.H. (eds): Modern and ancient continental shelf anoxia, Geol. Soc. Spec. Publ., 58, 311-334.

Littke, R. and ten Haven, H.L. (1989). Palaeoecologic trends and petroleum potential of Upper Carboniferous coal seams of western Germany as revealed by their petrographic and organic geochemical characteristics. Int. Journ. Coal, Geology, 13, 529-574.

Littke, R., Horsfield, B., and Leythaeuser, D. (1989). Hydrocarbon distribution in coals and dispersed organic matter of different maceral composition and maturities. Geol. Rundschau, 78, 391-410.

Littke, R., Klussmann, U., Krooss, B., and Leythaeuser, D. (1991d). Quantification of loss of calcite, pyrite, and organic matter due to weathering of Toarcian black shales and effects on kerogen and bitumen characteristics. Geochim. Cosmochim. Acta, 55, 3369-3378.

Littke, R., Leythaeuser, D., Radke, M., and Schaefer, R.G. (1990). Petroleum generation and migration in coal seams of the Carboniferous Ruhr basin, northwest Germany. In: Durand, B. and Béhar, F. (eds): Advances in organic geochemistry 1989, Org. Geochem., 16, 247-258.

Littke, R., Rotzal, H., Leythaeuser, D., and Baker, D.R. (1991a). Lower Toarcian Posidonia Shale in southern Germany (Schwäbische Alb) - organic facies, depositional environment, and maturity. Erdöl und Kohle, Erdgas, Petrochemie/Hydrocarbon Technology, 44, 407-414.

Littke, R. and Rullkötter J. (1987). Mikroskopische und makroskopische Unterschiede zwischen Profilen unreifen und reifen Posidonienschiefers aus der Hilsmulde. Facies, 17, 171-180.

Littke, R., Rullkötter, J., and Schaefer, R.G. (1991c). Organic and carbonate carbon accumulation on Broken Ridge and Ninetyeast Ridge, Central Indian Ocean. In: Weissel, J., Peirce, J., et al. (eds): Proceedings Ocean Drilling Program, Scientific Results 121, 467-487, College Station, Texas, (Ocean Drilling Program).

Littke, R., de Waal, S., and Biermanns, E. (1992). Pyrolysis results of artificial mixtures of Yallourn lignite with different minerals and of kerogen concentrates from these mixtures. Internal Report KFA/ICG-4 No. 500192.

Littke, R. and Welte, D.H. (1992). Hydrocarbon source rocks. In: Brown, Hawkesworth, and Wilson, C. (eds): Understanding the earth, 364-374, Cambridge University Press, Cambridge.

Loh, H., Maul, B., Prauss, M., and Riegel, W. (1986). Primary production, maceral formation and carbonate species in the Posidonia Shale of NW Germany. Mittl. Geol.-Paläont. Inst. Univ. Hamburg, 60, 397-421.

Lopatin, N.V. (1971). Temperature and geologic time as factors in coalification (in Russian). Akad. Nauk SSR Izv. Ser. Geol., 3, 95-106.

Lutz, M., Kaasschieter, J.P.H., and van Wijke, D.H. (1975). Geological factors controlling Rotliegend gas accumulations in the Mid-European basin. Proceed. 9th World Petroleum Congr., Applied Science Publ., II, 93-103.

Lyons, W.B., Hines, M.E. and Gaudette, H.E. (1984). Major and minor pore water geochemistry of modern marine sabkhas: The influence of cyanobacterial mats. In: Yehuda, C., Castenholz, R.W. and Halvorson, H.O. (eds): Microbial mats: Stromatolites, 411-423, A.R. Liss. Inc., New York

Lyons, P., Hatcher, P., Brown, F., Krasnow, M., and Larsen, R. (1985). Role of static load (overburden) pressure in coalification of bituminous and anthracitic coal. Proc. Int. Conf. Coal Science, Sidney, 620-623, Pergamon Press, Oxford.

Mackenzie, A.S. and McKenzie, D.P. (1983). Aromatization and isomerization of hydrocarbons in sedimentary basins formed by extension. Geol. Magazine, 120, 417-470.

Mackenzie, A.S., Price, I., Leythaeuser, D., Müller, P., Radke, M., and Schaefer, R.G. (1987). The expulsion of petroleum from Kimmeridge clay source rocks in the area of the Brae oil field, UK continental shelf. In: Brooks, J. and Glennie, K. (eds): Petroleum geology of northwest Europe, 865-877, Graham and Trotman, London.

Mahajan, O.P. and Walker, P.L., Jr. (1978). Porosity of coals and coal products. In: Karr, C., Jr. (ed.): Analytical methods for coal and coal products, 125-162, Academic Press, New York.

Manheim, F.T. (1961). A geochemical profile in the Baltic Cea. Geochim. Cosmochim. Acta, 25, 52-70.

Mann, U. (1987). Veränderung von Mineralmatrix und Porosität eines Erdölmuttergesteins durch einen Intrusivkörper. Facies, 17, 181-189.

Mann, U. and Müller, P.J. (1988). Source rock evaluation by well log analyses (Lower Toarcian, Hils Syncline). In: Mattavelli, L. and Novelli, L. (eds): Advances in organic geochemistry 1987, Org. Geochem. 13, 109-119.

Marzi, R. and Rullkötter, J. (1986). Organic matter accumulation and migrated hydrocarbons in deep sea sediments of the Mississippi fan and adjacent intraslope basins, northern Gulf of Mexico. Mitt. Geol.-Paläont. Inst. Univ. Hamburg, 60, 359-379.

Masters, J.A. (1984, ed.). Elmworth, case study of a deep basin gas field. Amer. Assoc. Petr. Geol. Memoir, 38, 316p.

Mc Limans, R.K. and Videtich, P.E. (1987). Reservoir diagenesis and oil migration: Middle Jurassic Oolite Limestone, Wealden basin, southern England. In: Brooks, J. and Glennie, K. (eds): Petroleum geology of northwest Europe, 119-128, Graham and Trotman, London.

McAuliffe, C. (1966). Solubility in water of paraffin, cycloparaffin, olefin, acetylene, cycloolefin and aromatic hydrocarbons. Journ. Phys. Chem., 70, 1267-1275.

McKenzie, D. (1978). Some remarks on the development of sedimentary basins. Earth Planet. Sci. Lett., 40, 25-32.

Mello, M.R., Telnaes, N., Gaglionone, P.C., Chicarelli, M.I., Brassell, S.C., and Maxwell, J.R. (1988). Organic geochemical characterisation of depositional palaeoenvironments of source rocks and oils in Brazilian margin basins. In: Mattavelli, L. and Novelli, L. (eds): Advances in organic geochemistry 1987, Org. Geochem., 13, 31-46.

Michelsen, J.R. and Kharvari Khorasani, G. (1990). Monitoring chemical alterations of individual oil-prone macerals by means of microscopical fluorescence spectrometry combined with multivariate data analysis. Org. Geochem., 15, 179-192.

Miller, R.G. (1989). Prediction of ancient coastal upwelling and related source rocks from palaeo-atmospheric pressure maps. Marine Petrol. Geol., 6, 277-283.

Moldowan, J., Albrecht, P., and Philp, R.P. (1992; eds.). Biological markers in sediments and petroleum. 411 p., Prentice Hall, Englewood Cliffs, New Jersey.

Moldowan, J.M., Sundararaman, P., and Schoell, M. (1986). Sensitivity of biomarker properties to depositional environment and/or source input in the Lower Toarcian of SW-Germany. In: Leythaeuser, D. and Rullkötter J. (eds): Advances in organic geochemistry 1985, Org. Geochem., 10, 915-926.

Morris, K. (1979). A classification of Jurassic marine shale sequences: An example from the Toarcian (Lower Jurassic) of Great Britain. Palaeogeography, Palaeoclimatology, Palaeoecology, 26, 117-126.

Morris, K.A. (1980). Comparison of major sequences of organic-rich mud deposits in the British Jurassic. Journ. Geol. Soc. London, 137, 157-170.

Moss, T.D., Riley, K.W., Saxby, J.D., Fookes, C.J.R. and Patterson, J.H. (1988). Effect of weathering of oil shale at Julia Creek (Australia) on kerogen, oil yields and oil properties. Fuel, 67, 1382-1385.

Mukhopadhyay, P.K., Rullkötter, J., and Welte, D.H. (1983). Facies and diagenesis of organic matter in sediments from the Brazil basin and the Rio Grande rise, deep sea drilling project, Leg 72. In: Barker, P.F., Carlson, R.L., and Johnson D.A. (eds): Init. Repts., DSDP, 72, p. 821-828.

Müller, G. and Blaschke, R. (1969). Zur Entstehung des Posidonienschiefers (Lias epsilon). Naturwissenschaften, 12, 635-636.

Müller, P.J. and Suess, E. (1979). Productivity, sedimentation rate and sedimentary organic matter in the oceans-organic carbon preservation. Deep-Sea Research, 27A, 1347-1362.

Naeser, N.D., Naeser, C.W., and McCulloh, T.H. (1989). The application of fission track dating to the depositional and thermal history of rocks in sedimentary basins. In: Naeser, N.D. and McCulloh, T.H. (eds): Thermal history of sedimentary basins, methods and case histories, 157-180, Springer-Verlag, New York.

Narr, W. and Burruss, R.C. (1984): Origin of reservoir fractures in Little Knife Field, North Dakota. Amer. Assoc. Petr. Geol. Bull., 68, 1087-1100.

Noble, R.A., Wu, C.H., and Atkinson, C.D. (1991). Petroleum generation and migration from Talang Akar coals and shales offshore N.W. Java, Indonesia. Org. Geochemistry, 17, 363-374.

Nöth, S. (1991). Die Conodontendiagenese als Inkohlungsparameter und ein Vergleich unterschiedlich sensitiver Diageneseindikatoren am Beispiel von Triassedimenten Nord- und Mitteldeutschlands. Boch. geol. geotechn. Arb., 37, 1-169.

Parrish, J.T. and Curtis, R.L. (1982). Atmospheric circulation, upwelling and organic-rich rocks in the Mesozoic and Cenozoic eras. Palaeogeogr., Palaeoclimatol.Palaeoecol., 40, 31-66.

Parrish, J.T., Ziegler, A.M., and Scotese, C.R. (1982). Rainfall patterns and the distribution of coals and evaporites in the Mesozoic and Cenozoic. Palaeogeogr., Palaeclimat., Palaeoecol., 40, 67-101.

Patnak, P. and Füchtbauer H. (1975). Temperature influencing the authigenic growth of silicates. IX. Int. Congr. Sedimentol. Abstr., Theme 7, 163-168.

Pedersen, T.F. and Calvert, S.E. (1990). Anoxia versus productivity: what controls the formation of organic-carbon-rich sediments and sedimentary rocks? Amer. Assoc. Petr. Geol. Bull., 74, 454-466.

Pedersen, T.F. and Calvert, S.E. (1991). Anoxia vs productivity: what controls the formation of organic-carbon-rich sediments and sedimentary rocks?: Reply. Amer. Assoc. Petr. Geol. Bull., 75, 500-501.

Peirce, J., Weissel, J., et al. (1989; eds). Proceedings of the Ocean Drilling Program, Initial Reports, 121, 1015p., College Station, Texas (Ocean Drilling Program).

Pelet, R. (1983). Preservation and alteration of present-day sedimentary organic matter. In: Bjoréy, M. (ed.): Advances in organic geochemistry, 241-250, John Wiley and Sons, Chichester.

Peters, K.E. (1986). Guidelines for evaluating petroleum source rocks using programmed pyrolysis. Amer. Assoc. Petr. Geol. Bull., 70, 318-329.

Philippi, G.T. (1975). The deep subsurface temperature controlled origin of the gaseous and gasoline-range hydrocarbons of petroleum. Geochim. Cosmochim. Acta, 39, 1353-1373.

Powell, T.G. (1986). Petroleum geochemistry and depositional setting of lacustrine source rocks. Mar. Petrol. Geol., 3, 200-219.

Powell, T.G., Boreham, C.J., Smyth, M., Russell, N., and Cook, A.C. (1991). Petroleum source rock assessment in non-marine sequences: pyrolysis and petrographic analysis of Australian coals and carbonaceous shales. Org. Geochemistry, 17, 375-394.

Pradier, B., Largeau, C., Derenne, S., Martinez, L., Bertrand, P., and Pouet, Y. (1990). Chemical basis of fluorescence alteration of crude oils and kerogens - I. Microfluorimetry of an oil and its isolated fractions; relationships with chemical structure. In: Durand, B. and Béhar, F. (eds): Advances in organic geochemistry 1989, Org. Geochem., 16, 451-460.

Prauss, M. and Riegel, W. (1989). Evidence from phytoplankton associations for causes of black shale formation in epicontinental seas. N. Jb. Geol. Paläont. Mh, 1989 (11), 671-682.

Prell W.L., Niitsuma N., et al. (1989; eds). Proceedings of the Ocean Drilling Program, Initial Reports 117, 1236p., College Station, Texas (Ocean Drilling Program).

Price, L.C. (1989). Primary petroleum migration from shales with oxygen-rich organic matter. Journ. Petr. Geol., 12, 289-324.

Püttmann, W., Eckardt, C.B., and Schaefer, R.G. (1988). Analysis of hydrocarbons in coal and rock samples by on-line combination of thermodesorption, gas chromatography and mass spectrometry. Chromatographia, 25, 279-287.

Quigley, T.M., Mackenzie, A.S., and Gray, J.R. (1987). Kinetic theory of petroleum generation. In: Doligez, B. (ed.): Migration of hydrocarbons in sedimentary basins, 649-666, Editions Technip, Paris.

Radke, M. (1987). Organic geochemistry of aromatic hydrocarbons. In: Brooks, J. and Welte, D.H. (eds): Advances in petroleum geochemistry, 2, 141-207, Academic Press, London.

Radke, M. (1988). Application of aromatic compounds as maturity indicators in source rocks and crude oils. Mar. Petr. Geol., 5, 224-236.

Radke, M., Leythaeuser, D., and Teichmüller, M. (1984). Relationship between rank and composition of aromatic hydrocarbons for coals of different origins. Org. Geochem., 6, 423-430.

Radke, M., Schaefer, R.G., Leythaeuser, D., and Teichmüller, M. (1980). Composition of soluble organic matter in coals: relation to rank and liptinite fluorescence. Geochim. Cosmochim. Acta, 44, 1787-1800.

Radke, M. and Welte, D.H. (1983). The methylphenanthrene index (MPI): A maturity parameter based on aromatic hydrocarbons. In: Bjorёy et al. (eds): Advances in organic geochemistry 1981, 504-512, John Wiley and Sons, Chichester.

Radke, M., Willsch, H., Leythaeuser, D., and Teichmüller, M. (1982). Aromatic components of coal: relation of distribution pattern to rank. Geochim. Cosmochim. Acta, 46, 1831-1848.

Radke, M., Willsch, H., and Teichmüller, M. (1990). Generation and distribution of aromatic hydrocarbons in coals of low rank. Org. Geochem., 15, 539-563.

Ramanampisoa, L., Radke, M., Schaefer, R.G., Littke, R., Rullkötter, J., and Horsfield, B. (1990). Organic-geochemical characterisation of sediments from the Sakoa coalfield, Madagascar. In: Durand, B. and Béhar, F. (eds): Advances in organic geochemistry 1989, Org. Geochem., 16, 235-243.

Reich, H. (1948). Geophysikalische Karte von Nordwest-Deutschland, 1:500,000, II. Gravimetrie. Celle/Hannover.

Reimers, C.E. and Suess, E. (1983). Spatial and temporal patterns of organic matter accumulation on the Peru continental margin. In: Thiede, J. and Suess, E. (eds): Coastal upwelling. Part B: Sedimentary records of ancient coastal upwelling, 311-346, Plenum Press, New York.

Rice, D.D., Clayton, J.L., and Pawlewicz, M.J. (1989). Characterization of coal-derived hydrocarbons and source-rock potential of coal beds, San Juan Basin, New Mexico and Colorado, U.S.A. In: Lyons, P.C. and Alpern, B. (eds). Coal: Classification, coalification, mineralogy, trace-element chemistry, and oil and gas potential, Int. Journ. Coal Geology, 13, 597-626.

Richter, D.K. (1985). Mikrodolomite in Crinoiden des Trochitenkalks (mo1) und die Wärmeanomalie von Vlotho. N. Jb. Geol. Paläont., Mh., 1985, 681-690.

Riegel, W., Loh, H., Maul, D., and Prauss, M. (1986). Effects and causes in a black shale event - the Toarcian Posidonia Shale of NW-Germany. Lect. Notes in Earth Sciences, 8, 267-276.

Riegraf, W. (1985). Mikrofauna, Biostratigraphie und Fazies im Unteren Toarcium Südwestdeutschlands und Vergleiche mit benachbarten Gebieten. Tübinger Mikropaläont. Mitteil., 3, 1-232.

Riegraf, W., Werner, G., and Lörcher, F. (1984). Der Posidonienschiefer, Ferd. Enke Verlag, Stuttgart.

Rightmire, C.T., Eddy, E.G., and Kirr, J.N. (1984, eds). Coalbed methane resources of the United States. Amer. Assoc. Petr. Geol. Studies in Geology, 17, 378p.

Riley, G.A. (1970). Particulate organic matter in seawater. Adv. Marine Biol., 8, 1-118.

Risk, M.J. and Rhoades, E.G. (1985). From mangroves to petroleum precursors. An example from tropical northeast Australia. Amer. Assoc. Petr. Geol. Bull., 69, 1230-1240.

Roedder, E. (1984). Fluid-inclusions. Rev. Miner., 12, 644p.

Ronov, A.B. (1958). Organic carbon in sedimentary rocks (in relation to the presence of petroleum). Translat. in Geochemistry, 5, 510-536.

Rotzal, H. (1990). Organisch-geochemische und organisch-petrologische Charakterisierung des organischen Materials im Posidonienschiefer (Lias epsilon) der Schwäbischen Alb, 111p., Diplomarbeit, RWTH Aachen.

du Rouchet , J. (1981). Stress fields, a key to oil migration. Amer. Assoc. Petr. Geol. Bull., 65, 74-85.

Rückheim, J. (1991). Reservoir-Geochemie. Ein neuer Wegweiser zur Erforschung von Erdöllagerstätten. Die Geowissenschaften, 9, 341-346.

Rullkötter, J., Flekken, P., and Welte, D.H. (1980). Organic petrography and extractable hydrocarbons of sediments from the Northern Phillippine sea, deep sea drilling project, Leg 58. In: de Vries Klein, G., Kobayashi, K., et al. (eds), Init. Repts., DSDP, 58, 755-762.

Rullkötter, J., Leythaeuser, D., Horsfield, B., Littke, R., Mann, U., Müller, P.J., Radke, M., Schaefer, R.G., Schenk, H.J., Schwochau, K., Witte, E.G., and Welte, D.H. (1988a). Organic matter maturation under the influence of a deep intrusive heat source: A natural experiment for quantitation of hydrocarbon generation and expulsion from a petroleum source rock. (Toarcian Shale, northern Germany). In: Mattavelli, L. and Novelli, L. (eds): Advances in organic geochemistry 1987. Org. Geochem., 13, 847-856.

Rullkötter, J., Littke, R., Hagedorn-Götz, I., and Jankowski, B. (1988b). Vorläufige Ergebnisse der organisch-geochemischen und organisch-petrographischen Untersuchungen an Kernproben des Messeler Ölschiefers. In: Franzen, J.L. and Michaelis, W. (eds): Der eozäne Messelsee - Eocene Lake Messel, Cour. Forsch-Inst. Senckenberg, 107, 37-52.

Rullkötter, J., Littke, R., and Schaefer, R.G. (1990). Characterization of organic matter in sulfur-rich lacustrine sediments of Miocene age (Nördlinger Ries, southern Germany). In: Orr, W.L. and White, C.M. (eds): Geochemistry of sulfur in fossil fuels, Amer. Chem. Soc. Symp. Series, 429, 149-169.

Rullkötter, J. and Marzi, R. (1988). Natural and artificial maturation of biological markers in a Toarcian shale from northern Germany. In: Mattavelli, L. and Novelli, L. (eds): Advances in organic geochemistry 1987, Org. Geochem., 13, 639-645.

Rullkötter, J. and Michaelis, W. (1990). The structure of kerogen and related materials: a review of recent progress and future trends. In: Durand, B. and Béhar, F. (eds): Advances in organic geochemistry 1989, Org. Geochem., 16, 829-852.

Sachsenhofer, R.F. (1990). Eine Inkohlungskarte des Steirischen Tertiärbeckens. Mitt. naturwiss. Ver. Steiermark, 120, 251-264.

Sachsenhofer, R.F. (1991). Maturität im Steirischen Tertiärbecken. Erdöl, Erdgas, Kohle, 107, 12-17.

Sachsenhofer, R.F. and Littke, R. (in press). Vergleich und Bewertung verschiedener Methoden zur Berechnung der Vitrinitreflexion am Beispiel von Bohrungen im Steirischen Tertiärbecken. Zentralbl. Geol. Paläont. I.

Savrda, C.E. and Bottjer, D.J. (1991). Oxygen-related biofacies in marine strata: an overview and update. In: Tyson, R.V. and Pearson, T.H. (eds): Modern and ancient continental shelf anoxia, Geol. Soc. Spec. Publ., 58, 201-220.

Schaefer, R.G. (1992). Zur Geochemie niedrigmolekularer Kohlenwasserstoffe im Posidonienschiefer der Hilsmulde. Erdöl u. Kohle, Erdgas Petroch./Hydrocarbon Techn., 45, 73-78.

Schaefer, R.G. and Littke, R. (1988). Maturity-related compositional changes in the low-molecular-weight hydrocarbon fraction of Toarcian shales. In: Mattavelli, L. and Novelli, L. (eds): Advances in organic geochemistry 1987, Org. Geochem., 13, 887-892.

Schaefer, R.G., Littke R., and Leythaeuser, D. (1991). Low-molecular-weight hydrocarbons in sediments of sites 752, 754, 755 (Broken Ridge), 757 and 758 (Ninetyeast Ridge), Central Indian Ocean. In: Weissel, J., Peirce, J., et al. (eds): Proceed. ODP, Sci. Results, 121, 457-466, College Station, TX (Ocean Drilling Program).

Scheidt, G. (1988). Ausbildung und Verteilung des dispersen organischen Materials im Ruhrkarbon. Boch. geol. geotechn. Arb., 28, 210 p.

Scheidt, G. and Littke, R. (1989). Comparative organic petrology of interlayered sandstones, siltstones, mudstones and coals in the Upper Carboniferous Ruhr basin, northwest Germany, and their thermal history and methane generation. Geol. Rundschau, 78, 375-390.

Scheihing, M.H. and Pfefferkorn, H.W. (1984). The taphonomy of landplants in the Orinoco delta: a model for the incorporation of plant parts in clastic sediments of Late Carboniferous age of Euramerica. Rev. Palaeobot. Palynol., 41, 205-240.

Schenk, H.J., Witte, E.G., Littke, R., and Schwochau, K. (1990). Structural modifications of vitrinite and alginite concentrates during pyrolytic maturation at different heating rates. In: Béhar, F. and Durand, B. (eds): Advances in organic geochemistry 1989, Org. Geochem., 16, 943-950.

Schmitz, H.H. (1980). Ölschiefer in Niedersachsen. Bericht der Naturhistorischen Gesellschaft Hannover, 123, 7-43.

Schramedei, R. (1991). Quantitative und qualitative Analyse von fossilen Kohlenwasserstoffen und ihrer Umwandlungsprodukte in Böden aus Posidonienschiefer (Ith-Hils-Mulde). Ber. Forschungszentrum Jülich, 2496, 1-153.

Schwab, F.L. (1976). Modern and ancient sedimentary basins: Comparative accumulation rates. Geology, 4, 723-727.

Schwarzkopf, T. and Leythaeuser, D. (1988). Oil generation and migration in the Gifhorn Trough, NW-Germany. In: Mattavelli, L. and Novelli, L. (eds): Advances in organic geochemistry 1987, Org. Geochem., 13, 245-254.

Secor, D.T. (1965). Role of fluid pressure on jointing. Amer. Journ. Science, 263, 633-646.

Seewald, H. (1982). Charkterisierung von Mikroporen und ihre Bedeutung für die Kohletechnik. Erdöl und Kohle, Erdgas, Petrochemie, 35, 418-427.

Seilacher, A. (1982). Ammonite shells as habitats in the Posidonia shales of Holzmaden - floats or benthic island? N. Jb. Geol. Paläont. Mh., 1982, 98-114.

Shibaoka, M. and Smyth, M. (1975). Coal petrology and the formation of coal seams in some Australian sedimentary basins. Econ. Geol., 70, 1463-1473.

Shimkus, K.M. and Trimonis, E.S. (1974). Modern sedimentation in Black Sea. In: Degens, E.T. and Ross, D.A. (eds): The Black Sea - geology, chemistry, and biology. Amer. Assoc. Petr. Geol. Memoir, 20, 249-278.

Smith, A.H.V. (1957). The sequence of miospore assemblages associated with the occurrence of crassidurite in coal seams of Yorkshire. Geol. Mag., 93, 345-363.

Smith, A.H.V. (1962). The paleoecology of Carboniferous peats based on the miospores and petrography of bituminous coals. Proc. Yorkshire Geol. Soc., 33, 423-478.

Smyth, M. (1989). Organic petrology and clastic depositional environments with special reference to Australian coal basins. Intern. Journ. Coal. Geol., 17, 635-656.

Snowdon, L.R. and Powell, T.G. (1982). Immature oil and condensate-modification of hydrocarbon generation model for terrestrial organic matter. Amer. Assoc. Petr. Geol. Bull., 66, 775-788.

Solomon, P.R. and Miknis, F.P. (1980). Use of Fourier transform infrared spectroscopy for determining oil shale properties. Fuel, 59, 893-896.

Spaulding, S. (1991). Neogene nannofossil biostratigraphy of Sites 723 through 730, Oman continental margin, northwestern Arabian Sea. In: Prell, W.L., Niitsuma, N., et al. (eds): Proceedings of the Ocean Drilling Program, Sci. Results, 117, 5-36.

Stach, E. (1982). The macerals of coal. In: Stach, E., et al. (eds): Stach's textbook of coal petrology, 87-139, Gebr. Bornträger, Berlin.

Stach, E., Mackowsky, M.Th., Teichmüller, M., Taylor, G.H., Chandra, D., and Teichmüller, R. (1982). Stach's textbook of coal petrology, 535 p., Gebrüder Bornträger, Stuttgart.

Stainforth, J.G. and Reinders, J.E.A. (1990). Primary migration of hydrocarbons by diffusion through organic matter networks, and its effect on oil and gas generation. In: Durand, B. and Béhar, F. (eds): Advances in organic geochemistry 1989, Org. Geochem., 16, 61-74.

Staplin, F.L. (1977). Interpretation of thermal history from color of particulate organic matter - a review. Palynology, 1, 9-18.

Stein, R. (1986). Surface-water paleo-productivity as inferred from sediments deposited in oxic and anoxic deep water environments of the Mesozoic Atlantic ocean. Mitt. Geol.-Paläont. Inst. Univ. Hamburg, 60, 55-70.

Stein, R. (1991). Accumulation of organic carbon in marine sediments. Lect. Notes Earth Sciences, 34, 1-217.

Stein, R., ten Haven, H.L., Littke, R., Rullkötter, J. and Welte, D.H. (1989). Accumulation of marine and terrigenous organic carbon at upwelling Site 658 and nonupwelling Sites 657 and 659: implications for the reconstruction of paleoenvironments in the eastern subtropical Atlantic through Late Cenozoic times. In: Ruddiman, W., Sarnthein, M., et al. (eds): Proceedings of the Ocean Drilling Program, Sci. Results, 108, 361-385.

Stein, R. and Littke, R. (1990). Organic-carbon-rich sediments and palaeoenvironment: results from Baffin Bay (ODP-Leg 105) and the upwelling area off northwest Africa (ODP-Leg 108). In: Huc, A.Y. (ed.): Deposition of Organic Facies, Amer. Assoc. Petr. Geol. Studies in Geology, 30, 41-56.

Stein, R., Rullkötter, J., Littke, R., Schaefer, R.G., and Welte, D.H. (1988). Organofacies reconstruction and lipid geochemistry of sediments from the Galicia margin, Northeast Atlantic (ODP Leg 103). In: Boillot, G., Winterer E.L., et al. (eds): Proceedings of the Ocean Drilling Programm, Sci. Results, 103, 567-585.

Styan, W.B. and Bustin, R.M. (1983). Petrography of some Fraser river delta peat deposits: coal maceral and microlithotype precursors in temperate-climate peats. Int. Journ. Coal, Geol., 2, 321-370.

Suess, E., von Huene, R., et al. (1988; eds). Proceedings of the Ocean Drilling Program, Init. Repts, 112, 1015p.

Suess, E. and Thiede, J. (1983; eds). Coastal upwelling. Part A: Responses of the sedimentary regime to present coastal upwelling, 610p., Plenum Press, New York.

Summerhayes, C.P. (1983). Sedimentation of organic matter in upwelling regimes. In: Thiede, J. and Suess, E. (eds): Coastal upwelling. Part B: Sedimentary records of ancient coastal upwelling, 29-72, Plenum Press, New York.

Sundararaman, P., Schoell, M., Littke, R., Baker, D.R. Leythaeuser, D., and Rullkötter, J. (1991). Depositional environment of Toarcian shales from NW-Germany as monitored by porphyrins and carbon isotopes. In: Manning, D.A.C. (ed.): Organic Geochemistry. Advances and applications in energy and the natural environment, 389-390, Manchester Univ. Press, Manchester.

Swanson, V.E., and Palacas, J.G. (1965). Humate in coastal sands of northwest Florida. U.S. Geol. Surv., Bull, 1214-B.

Sweeney, J.J. and Burnham, A.K. (1990). Evaluation of a simple model of vitrinite reflectance based on chemical kinetics. Amer. Assoc. Petr. Geol. Bull., 74, 1559-1570.

Teichmüller, M. (1974). Über neue Macerale der Liptinit-Gruppe und die Entstehung des Micrinits. Fortschr. Geol. Rheinld. Westf., 24, 37-64.

Teichmüller, M. (1982). Origin of the petrographic constituents of coals. In: Stach, E., et al. (eds): Stach's textbook of coal petrology, 219-294, Gebrüder Bornträger, Stuttgart.

Teichmüller, M. (1986). Organic petrology of source rocks, history and state of the art. In: Leythaeuser, D. and Rullkötter, J. (eds): Advances in organic geochemistry 1985, Org. Geochem., 10, 581-599.

Teichmüller, M. and Ottenjann, K. (1977). Liptinite und lipoide Stoffe in einem Erdölmuttergestein. Erdöl und Kohle, 30, 387-398.

Teichmüller, M. and Teichmüller, R. (1982). The geological basis of coal formation. In: Stach, E., et al., (eds): Stach's Textbook of coal petrology, 5-86, Gebr. Bornträger, Berlin.

Teichmüller, M., Teichmüller, R., and Bartenstein, H. (1984). Inkohlung und Erdgas - eine neue Inkohlungskarte der Karbon-Oberfläche in Nordwestdeutschland. Fortschr. Geol. Rheinld. Westf., 32, 11-34.

Thiede, J. and Suess, E. (1983; eds). Coastal upwelling. Part B: Sedimentary records of ancient coastal upwelling, 610p., Plenum Press, New York.

Thimons, B. and Kissell, F.N. (1973). Diffusion of methane through coal. Fuel, 52, 274-280.

Thomas, B.M. (1982). Land plant source rocks for oil and their significance in Australian basins. Austr. Petr. Expl. Assoc. Journ., 22, 166-178.

Thompson, K.F.M. (1979). Light hydrocarbons in subsurface sediments. Geochim. Cosmochim. Acta, 43, 657-672.

Thompson, S., Cooper, B.S., Morley, R.J., and Barnard, P. (1985). Oil-generating coals. In: Thomas, B.M., et al. (eds): Petroleum geochemistry in exploration of the Norwegian Shelf, 59-73 Graham and Trotman, London.

Thunell, R.C., Williams, D.F., and Belyea, P.R. (1984). Anoxic events in the Mediterranean sea in relation to the evolution of late Neogene climates. Marine Geol., 59, 105-134.

Ting, F.T.C. (1982). Coal macerals. In: Brooks, J. (ed.): Organic maturation studies and fossil fuel exploration, 379-392, Academic Press, Orlando.

Tissot, B.P., Califet-Debyser, Y., Deroo, G., Oudin, J.L. (1971). Origin and evolution of hydrocarbons in Early Toarcian Shales. Amer. Assoc. Petr. Geol. Bull., 55, 2177-2193.

Tissot, B.P. and Espitalié, J. (1975). L' évolution thermique de la matière organique des sédiments: Applications d' une simulation mathématique. Rev. Inst. Franç. Pétr., 30, 743-777.

Tissot, B.P. and Welte, D.H. (1984). Petroleum formation and occurrence. 2nd edition, 699p., Springer, Berlin.

Tissot, B.P., Welte, D.H., and Durand, B. (1987). The role of geochemistry in exploration, risk evaluation and decision making. Proceed. 12th World Petr. Congr., 2, 99-112.

Tourtelot, H.A. (1979). Black shale - its deposition and diagenesis. Clays Clay Minerals, 27, 313-321.

Tyson, R.V. (1987). The genesis and palynofacies characteristics of marine petroleum source rocks. In: Brooks, J. and Fleet, A.J. (eds): Marine petroleum source rocks, Geol. Soc. Spec. Publ., 26, 47-67.

Tyson, R.V. and Pearson, T.H. (1991; eds). Modern and ancient continental shelf anoxia. Geol. Soc. Spec. Publ., 58, 470p.

Ungerer, P. (1990). State of the art of research in kinetic modelling of oil formation and expulsion. In: Durand, B. and Béhar, F. (eds): Advances in organic geochemistry 1989, Org. Geochem., 16, 1-25.

Ungerer, P., Béhar, F., Villalba, M., Heum, O.R., and Audibert, A. (1988). Kinetic modelling of oil cracking. In: Mattavelli, L. and Novelli, L. (eds): Advances in organic geochemistry 1987, Org. Geochem., 13, 857-868.

Ungerer, P., Bessis, F., Chenet, P.Y., Durand, B., Nogaret, E., Chiarelli, A., Oudin, J., and Perrin, J.F. (1984). Geological and geochemical models in oil exploration: principles and practical examples. In: Demaison, G. (ed.): Petroleum geochemistry and basin evolution, Amer. Assoc. Petr. Geol. Mem., 35, 53-77.

Urlichs, M. (1977). The Lower Jurassic in southwestern Germany. Stuttgarter Beiträge zur Naturkunde, Reihe B., 24, 1-30.

Valeton, I. (1988). Verwitterung und Verwitterungslagerstätten. In: Füchtbauer, H. (ed.): Sedimente und Sedimentgesteine, 11-68, E. Schweizerbart, Stuttgart.

Volkmann, J.K., Allen, D.I., Stevenson, P.L., Burton, H.R. (1986). Bacterial and algal hydrocarbons in sediments from a saline Antarctic lake, Ace Lake. In: Leythaeuser, D. and Rullkötter, J. (eds): Advances in organic geochemistry 1985, Org. Geochem., 10, 671-682.

Waples, D.W. (1980). Time and temperature in petroleum formation: Application of Lopatin's method to petroleum exploration. Amer. Assoc. Petr. Geol. Bull., 64, 916-926.

Waples, D.W., Kamata, H., and Suizu, M. (1992a). The art of maturity modeling. Part 1: Finding a satisfactory geologic model. Amer. Assoc. Petr. Geol. Bull., 76, 31-46.

Waples, D.W., Suizu, M., and Kamatra, H. (1992b). The art of maturity modeling. Part 2: Alternative models and sensitivity analysis. Amer. Assoc. Petr. Geol. Bull., 76, 47-66.

Watanabe, T., Langseth, M.G., and Anderson, R.N. (1977). Heat flow in back-arc basins of the western Pacific. In: Talwani, M. and Pitmann, W.C. (eds): Island arcs, deep sea trenches and back-arc basins, 137-161, Washington (American Geophys. Union).

Weedon, G. P. (1986). Hemipelagic shelf sedimentation and climatic cycles: the basal Jurassic (Blue Lias) of South Britain. Earth Planet. Sci. Letters, 76, 321-335.

Wefer, G., Heinze, P., and Suess, E. (1990). Stratigraphy and sedimentation rates from oxygen isotope composition, organic carbon content and grain-size distribution at the Peru upwelling region: Holes 680B and 686B. In: Suess, E., von Huene, R., et al. (eds): Proceedings of the Ocean Drilling Program, Sci. Results, 112, 355-368.

Wehner, H., Gerling, P., Hiltmann, W., and Kockel, F. (1989). Erdöl-Charakteristik und Öl-Muttergestein-Korrelation im Niedersächsischen Becken. Nachr. Deutsche Geol. Ges., 41, 77-78.

Weissel, J., Pierce, J., et al. (1991, eds.). Proceedings of the Ocean Drilling Program, Sci. Results, 121, 990p.

Welte, D.H. (1979). Organisch-geochemische Untersuchungen zur Bildung von Erdöl-Kohlenwasserstoffen an Gesteinen des mittleren Oberrhein-Grabens. Fortschr. Geol. Rheinld. Westf., 27, 51-74.

Welte, D.H. and Waples, D.W. (1973). Über die Bevorzugung gradzahliger n-Alkane in Sedimentgesteinen. Naturwiss., 60, 516.

Welte, D.H. and Yükler, A. (1981). Petroleum origin and accumulation in basin evolution. Amer. Assoc. Petr. Geol. Bull., 65, 1387-1396.

Wenger, L.M. and Baker, D.R. (1986). Variations in organic geochemistry of anoxic-oxic black shale-carbonate sequences in the Pennsylvanian of the Midcontinent, USA. In: Leythaeuser, D. and Rullkötter, J. (eds): Advances in organic geochemistry 1985, Org. Geochem., 10, 85-95.

Witte, E.G., Schenk, H.J., Müller, P.J., and Schwochau, K. (1988). Structural modification of kerogen during natural evolution as derived from 13C CP/MAS NMR, IR spectroscopy and Rock Eval pyrolysis of Toarcian shales. In: Mattavelli, L. and Novelli, L. (eds): Advances in organic geochemistry 1987, Org. Geochem., 13, 1039-1044.

Woldstedt, P. (1958). Das Eiszeitalter; Grundlinien einer Geologie des Quartärs. 2. Europa, Vorderasien und Nordafrika im Eiszeitalter. 438p. Enke Verlag, Stuttgart.

Wygrala, B.P. (1989). Integrated study of an oil field in the southern Po basin, northern Italy. Ber. KFA Jülich, 2313, 1-313.

Yalcin, M.N. and Welte, D.H. (1989). The thermal evolution of sedimentary basins and significance for hydrocarbon generation. Bull. Turkish Petr. Geol., 1, 12-26.

Ziegler, P.A. (1982). Geological atlas of western and central Europe, Elsevier, Amsterdam.

van der Zwaan, G.J. and Jorissen, F.J. (1991). Biofacial pattern in river-induced shelf anoxia. In: Tyson, R.V. and Pearson, T.H. (eds): Modern and ancient continental shelf anoxia, Geol. Soc. Spec. Publ., 58, 65-82.

Lecture Notes in Earth Sciences

Vol. 43: N. Clauer, S. Chaudhuri (Eds.), Isotopic Signatures and Sedimentary Records. VIII, 529 pages. 1992.

Vol. 44: D. A. Edwards, Turbidity Currents: Dynamics, Deposits and Reversals. XIII, 175 pages. 1993.

Vol. 45: A. G. Herrmann, B. Knipping, Waste Disposal and Evaporites. XII, 193 pages. 1993.

Vol. 47: R. L. Littke, Deposition, Diagenesis and Weathering of Organic Matter-Rich Sediments. IX, 216 pages. 1993.

Vol. 48: B. R. Roberts, Water Management in Desert Environments. XVII, 337 pages. 1993.